The Secret Science of Superhei

The Secret Science of Superheroes

Edited by

Mark Lorch
University of Hull, UK
Email: m.lorch@hull.ac.uk

and

Andy Miah
University of Salford, UK
Email: a.miah@salford.ac.uk

THE QUEEN'S AWARDS
FOR ENTERPRISE:
INTERNATIONAL TRADE
2013

Print ISBN: 978-1-78262-487-5
EPUB ISBN: 978-1-78801-265-2

A catalogue record for this book is available from the British Library

The Royal Society of Chemistry is a charity, registered in England and Wales, Number 207890, and a company incorporated in England by Royal Charter (Registered No. RC000524), registered office: Burlington House, Piccadilly, London W1J 0BA, UK, Telephone: +44 (0) 207 4378 6556.

Visit our website at www.rsc.org/books

Printed in the United Kingdom by CPI Group (UK) Ltd, Croydon, CR0 4YY, UK

Preface

If you are going to enjoy a superhero movie you need to be able to suspend disbelief, especially if you are scientifically inclined. There is just too much that is just plain impossible. If we moaned about every little detail that wasn't quite correct by some law of physics, then we'd ruin the film for almost anyone who has the misfortune of being within shouting distance. This is why *The Secret Science of Superheroes* doesn't try to pick holes in movies or comic book stories. We don't spend time explaining why spaceships don't need wings, or why starships always have the same orientation when they meet in space. We don't quibble over the fact that lasers beams can't be seen from the side, unless there is something around to scatter the light, or why they don't make '*puchu puchu*' noises when fired. Neither do we point out that Iron Man's armour is no use to him in a crash, since what he requires is something more like a crumple zone or an air bag, instead of a suit constructed from inflexible titanium. We don't discuss why being hit by a bullet won't throw you backwards, because of something we call momentum.

These are all great examples of reality being suspended for the sake of drama and we're cool with that, because good movies rely on presenting the impossible, even if the improbable is frowned upon. So we are fine with faster than light travel, fiery explosions in space (no oxygen = no fire), and laser sound effects. However, indestructible metals, web-slinging humans and invisibility leave us pondering how science might explain them.

The Secret Science of Superheroes
Edited by Mark Lorch and Andy Miah
© The Royal Society of Chemistry, 2017
Published by the Royal Society of Chemistry, www.rsc.org

So, this book is about trying to suspend the improbable. It is about the missing scenes (and the missing science) that could be in movies and comics if what actually were shown to us had some scientifically feasible explanation.

Now, we could have taken a typical solitary, leisurely approach to penning this book, holed up in an office writing over months and years. But if we've learnt anything from superhero flicks it's that all the best stories have teams: X-Men, The Justice League and the Fantastic Four trump the lonely Spider-Man or Batman any day. Secondly, faster is better. You never hear of a hero travelling slower than a plodding tortoise or proclaiming to be the most ponderous man alive.

No, a book about heroes needs a more rapid fire, heroic approach. Which is why we assembled a league of extraordinary scientists and set them the Herculean task of writing this book in just 36 hours. Plonked in the middle of the Manchester Science Festival and Salford University's Science Jam, in a blur of flying fingers worthy of the Flash we cranked out over 200 pages. We delved into all the nitty gritty science that fascinates us but seems to have been overlooked by movie makers.

During our frenetic weekend of typing – punctuated with regular trips down rabbit holes, comic strips out of context caused much mirth, google it – a means of charting superpowers emerged. Our diagram categorizes powers depending on whether they are passive (are they working all the time, like The Thing's armoured skin) or active (they need to be invoked such as Spider-Man's web slingers). Another dimension to the chart indicates whether the power is intrinsic to the hero (*i.e.* it can't be removed) or is extrinsic (meaning it's something that is associated with them, think of Captain America's shield). Finally, the reach of the power is indicated from the distance from the centre of the chart. So something that only affects the hero themselves appears in the middle, whilst a power that can affect something anywhere on the globe (or beyond) appears on the edge. The superhero, intrinsic, extrinsic, location diagram (otherwise known as The SHIELD) also turned out to be a rather neat alternative to the conventional contents page.

As a collection, the book covers everything from the intricacies of individual superpowers and what they might require of the real world in order to be possible, while also examining where examples of superhumanness are found within the capacities of other species. Together, the essays show how superheroes are born out of a place where we have a remarkable number of natural precedents, even if their use of these powers goes far beyond what is expected of the average individual. In this respect, it seems more the super-virtues of the heroes rather than simply their functional capacities which elevates them beyond simply super-ness into a world of heroic deeds. It may turn out that we can also find a way to design superhuman virtues to make us more heroic, but this would be the subject of another book.

Before we get stuck in, a few thanks are necessary, as this book relied on some great support. First, thank you to the University of Salford and the team at Media City, for its hosting of the Book Sprint and bringing together authors to think through ideas and work through words. One of the reasons for why we do this is to re-think the way in which edited volumes are created and the sprint allowed us to forge new relationships which we hope will last. We are also grateful to Manchester Science Festival, which serves as the inspiration for this project. Thanks also to the illustrators who accompanied us on the weekend, namely Syeda Khanum and Romica Spiegl-Jones. In the end we are sorry that your delightful images didn't make it into the book, but your input and inspiration was very much appreciated. A special thanks to Andy Brunning for the fabulous infographics that appear throughout the book; be sure to check out his incredible chemistry graphics at http://www.compoundchem.com/. And a final thanks to The Conversation (www.conversation.com); most of the authors were found by searching through their fabulous back-catalogue of writers.

Mark Lorch and Andy Miah

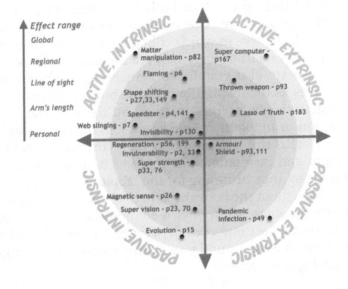

THE SUPERHERO INTRINSIC EXTRINSIC LOCATION DIAGRAM

Effect range

Global

Regional

Line of sight

Arm's length

Personal

ACTIVE, INTRINSIC

ACTIVE, EXTRINSIC

PASSIVE, INTRINSIC

PASSIVE, EXTRINSIC

Matter manipulation - p82

Super computer - p167

Flaming - p6

Thrown weapon - p93

Shape shifting - p27,33,149

Speedster - p4,141

Lasso of Truth - p183

Web slinging - p7

Invisibility - p130

Regeneration - p56, 199

Invulnerability - p2, 33

Armour/Shield - p93,111

Super strength - p33, 76

Magnetic sense - p26

Super vision - p23, 70

Pandemic infection - p49

Evolution - p15

Contents

The Secret Science of Superheroes
Edited by Mark Lorch and Andy Miah
© The Royal Society of Chemistry, 2017
Published by the Royal Society of Chemistry, www.rsc.org

The Breakfasts of Superheroes

MARK LORCH

School of Mathematics and Physical Science, University of Hull, HU6 7RX, UK
E-mail: M.lorch@hull.ac.uk

1.1 INTRODUCTION

One of the most obvious but overlooked questions that surrounds the science of being a superhero concerns the nutritional needs that would be required to have the capacity of superhuman powers. After all, a handful of oat cakes for Peter Parker (AKA Spider-Man), as he dashes off on a day of acrobatic crime fighting across Metropolis, just aren't going to cut the mustard. There has to be something else going on. And Peter Parker knows it. He must know that his daring deeds are only possible if he is fuelled by a proper diet. The oat cake breakfast may just be part of his cover story, but Spider-Man's breakfast is just the tip of the iceberg. After all, if Olympic athletes need a support team inspecting every morsel that passes their lips, then there must be an army of nutritionists monitoring the dietary requirements of the great crime fighting squads. It may not crop up in the comics

The Secret Science of Superheroes
Edited by Mark Lorch and Andy Miah
© The Royal Society of Chemistry, 2017
Published by the Royal Society of Chemistry, www.rsc.org

and movies, but I bet there's a course entitled '*Nutrition for the Gifted – 101*' on the curriculum at hero schools. Education about healthy eating can't start early enough.

1.2   FOOD FOR ALL

Before we get stuck into the nutritional needs of the superpowered, let's remind ourselves of what mere mortals in this reality tuck into. Mrs Average and Mr Ordinary need about 2000–2500 Calories[†] per day.[1] Now, compare that to the most athletic amongst us (Table 1.1). Many professional athletes feed on 3000–4000 Calories per day. Whilst some Olympic swimmers, at the height of their training regimes, claim to increase their diet from an already hefty 5000–6000 Calories to a stomach stretching 10 000 Calories each day![2]

The difference between the athletes' diet and those of the average Jo and Joe is subtler than just energy intakes; there's also a significant change in nutritional balance. The calories fuel an athlete's daily activities, but the machinery that powers their achievements is muscle, which is made of protein. Consequently, the athletes must increase their protein intake by as much as four times over that of Mr Ordinary. In short, they change their diet to take into account their activities, providing them with the nutrients to build a body prepared for the task ahead.

Just as athletes tweak their diets to feed the demands of their sport, superheroes must take into account their body's needs. Of course, many of our superheroes act in ways akin to real world athletes and their diets might well be very similar to Olympians. But there are some unusual cases who will have extraordinary nutritional needs to go along with their strange powers. This may be the result of the massive amount of energy they expel, but in other cases heroes need to remember that, with great powers come great side-effects.

1.3   A SUPER SIDE-EFFECT

Consider poor old Ben Grimm, also known as The Thing, a man trapped inside a craggy orange hide. His powers and stone exoskeleton might afford him fabulous stamina, strength and

---

[†]A nutritional or food 'Calorie' is equivalent to 1000 'scientific' calories. This means that a food Calorie is the same as a kcal, which is what is used in Table 1.1.

Table 1.1   Daily diet of humans.

	Mrs Average	Mr Ordinary	Pro-Rugby guy	Mr Olympic gymnast	Tennis-man	Pool-man
Energy (kcal)	2000	2500	3000	3800	3100	5800
Protein (g)	45	55	230	147	255	280
Carbohydrate (g)	230	300	300	720	420	490
Sugar (g)	90	120	65	210	210	71
Fat (g)	70	90	122	44	64	300

resistance to injury, but at the cost of a monstrous form. There is also something else to take into account ... His stony exterior makes him impervious to ultraviolet radiation. There's no tanning or sunburn for The Thing. Well, that's not such a terrible side-effect, you might think. After all, if you're a big orange stony hulk (with a small H) the least of your worries will be an inability to soak up the rays whilst sunning yourself on a Caribbean vacation. However, without exposure to UV, Ben will suffer from a vitamin deficiency that results in fatigue and muscle weakness and he will even have trouble thinking – all things a superhero really should do his best to avoid.

Most vitamins are synthesized biologically through a series of enzyme-controlled reactions that start with basic building blocks and, by stitching them together here and nicking a bond there, the final vitamin is formed. Sometimes an organism might be missing enzymes that allows it to make a vitamin, but that's generally OK because some other beast, bug or plant that we consume will have made it, transferring it into us when we eat. A perfect example of super-failing in humans (along with guinea pigs, capybaras and bats) is vitamin C, which we really need, but can't generate. So instead we need to get it from our food.

However, vitamin D is rather unusual. Nowhere on the planet is there an organism that can make it without external help; they all require energy from the sun to smash open a ring in a molecule called 7-dehydrocholesterol (Figure 1.1). This process frees up the new compound to twist (in a process called isomerization) into the shape of vitamin D, and this all goes on in your skin when you are out in the sun. So, if you have the misfortune of being covered in orange stone, then there is no way the sunlight will reach the 7-dehydrocholesterol. Instead you would need to ensure your diet includes plenty of the required vitamin.

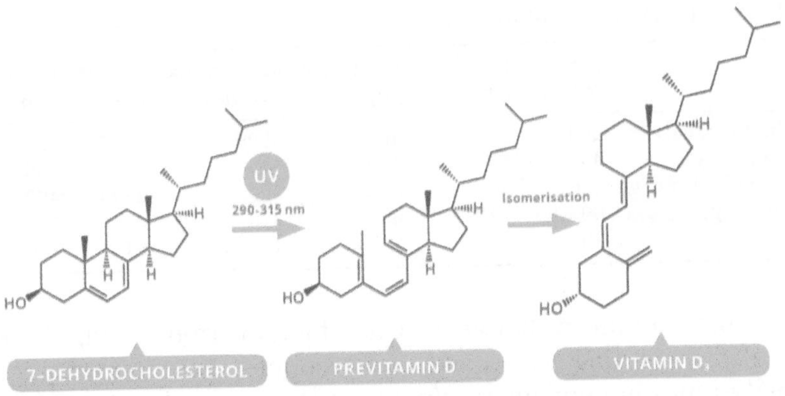

Figure 1.1 Our bodies can't quite finish making vitamin D by themselves; the final step needs an energetic 'kick' from ultraviolet light. This breaks a bond in one of the rings of a cholesterol-like molecule, allowing the ring to unravel, forming vitamin D. © Andy Brunning 2017.

Spend plenty of time outside and you'll probably manufacture all the vitamin D you need (assuming you don't cover yourself up). However, the recommended daily allowance assumes you generally lock yourself away in a dark room (reading comic books?), live in northern latitudes (where the sun isn't so strong), and cover yourself in factor 50 sun cream. So, the suggestion is that you should eat 15 micrograms of vitamin D each day. Ben probably needs to take in this amount as well, which means that a breakfast of oily fish[‡] (kippers maybe?) and he will be just fine.

1.4 FAST FOOD AND FLASH DIETS

Beyond simple strength and energy, superheroes who are super speedy also have unique nutritional needs in the morning. Let's take Marvel's Quicksilver as a start. In his original incarnation he maxed out at the speed of sound, although since then his powers have grown. From recent footage some clever clogs has calculated Quicksilver can whiz along at well over ten times the speed of sound (12 000 kph) over short distances.[3] But, let's keep things realistic and work out what he might consume at breakfast to

---

[‡]Fish eat plankton and the plankton float around near the surface of water soaking up the rays, which is how the vitamin D is transferred.

power a 30 minute Mach 1 run?[§] Luckily, sport scientists have a handy equation for calculating energy usage for runners,[4] so what happens if we apply it to our supersonic hero?

First, we need to work out how much oxygen our heroic athlete uses. By hooking up regular athletes to monitors whilst they are resting, we know that they generally use 3.5 mL of oxygen per minute per kg of body mass. As we all know, once we start running we need more oxygen and start breathing harder, so we need take this into account as well. The volume of oxygen used in a minute is known as VO_2 and is calculated as follows:

$VO_2 = 3.5 + (0.2 \times$ speed in meters per minute) when running on the flat[¶]

So, our runner speeding along at the speed of sound (20 400 meters per minute) uses:

$$VO_2 = 3.5 + (0.2 \times 20\,400) = 4083.5 \text{ mL kg}^{-1} \text{ min}^{-1}$$

According to Marvel Directories, Quicksilver weighs 80 kg (175 lb).[5] So, on his (quick) morning jog, he'd need 327 litres of oxygen per minute (80 kg × 4083.5 mL kg^{-1} min^{-1} = 327 L min^{-1}). Because air is about 20% oxygen, that means he's breathing in 1633 litres of air every 60 seconds![‖]

To utilize the oxygen he needs some chemical energy, which comes from food, and we know that works out at 5 kcal per litre of oxygen used. So we have another bit of maths:

$$5 \text{ kcal} \times 327 \text{ L min}^{-1} = 1635 \text{ kcal min}^{-1}$$

This gives Quicksilver needing 49 050 kcal for his half-hour morning run. If you look back at Table 1.1, you may notice that this is much more than you'd eat in a day, but how much more?

Peanut butter and jelly (jam) sandwiches are a favourite in the US (the land of our hero) so maybe he's partial to them for breakfast? They come in at about 500 kcal per round, so before taking off on his run Quicksilver might want to tuck into about 100 of them!

[§]We're going to assume he's doing his morning workout on the treadmill, which means we don't have to take wind resistance into account.

[¶]For dashes up slopes you'd need an extra function: $VO_2 = 3.5 + (0.2 \times$ speed) + (0.9 × speed × gradient).

[‖]A side effect of this is that Quicksilver could disable his opponents by running around in circles. He'd rapidly use all the oxygen in the room, so suffocating anyone in there with him.

1.5   FEEDING THE FLAMES

Super speed isn't the only power that's going to require a calorie-packed diet. In fact, compared to some of the superpowers that are out there, it may even be one of the more modestly energy-needy powers. In comparison, pyrokinetic superheroes, like Johnny Storm, who bursts into flames at will, generate huge amounts of heat, which certainly requires loads of energy. How Johnny ignites himself isn't entirely clear. On the one hand, he seems to need oxygen to maintain his flames, which suggests they are the result of combustion. However, comic book lore consistently states that The Human Torch's hot plumes are actually plasma. Now, plasma is a gas-like matter created by stripping electrons from their nuclei, like you get in stars. But there's no good reason why generating plasma needs oxygen. Actually, the mechanism by which he 'flames-on' doesn't really matter. Either way, he needs to heat up and then maintain the desired temperature, so how does he do it?

Johnny's energy requirements can be broken into two steps. First, how much energy is required to heat him from a normal 37 °C to his minimum 'flame-on' temperature of 416 °C?[6] And then, what's required to maintain that temperature?** The energy, in calories (small 'c' calories this time) needed (E) to heat up is simple enough and is given by the equation:

$$E = c \times m \times \Delta T$$

where m is Johnny's mass (77 kg) and ΔT is the change in temperature (379 °C, in this case, the difference between the target of 416 °C and normal body temperature). C is something called 'heat capacity', which is how much energy is needed to increase a kilogram of a given type of matter by 1 °C. (I'll use water's heat capacity since humans are basically bags of the stuff.) Conveniently this is 1 kcal kg^{-1} K^{-1}†† (because that's how a calorie

---

** I'm not the first person to work all this out. A student paper in *Physics Special Topics*[7] laid out these steps and calculated the Human Torch's energy requirements (although they did make an error which means their result is out by a factor of 10). The authors assumed Storm is making hydrogen plasma following the physics of our universe, which needs to be at a temperature of 10 000 degrees centigrade. I'm sticking to the temperatures stipulated in the comics.

†† K being the temperature unit Kelvin. 1 degree Kelvin is the same size as 1 degree centigrade, so it doesn't really matter which unit you use to define a calorie, but the convention is to use Kelvin.

is defined). Plugging all these values into the equation we get a figure of 29 183 kcal. That's just to heat him up.

Next up, how to stay hot. As anyone who pays a heating bill knows you've got to pump in energy to keep something warm. Flaming superheroes will have to deal with this same cooling issue. The rate of heat loss through cooling was worked our centuries ago by a real superhero scientist, Sir Isaac Newton.[8] His equation for energy lost by convection in 1 second is:

$$E = h \times A \times (T_{hot} - T_{cold})$$

E is the heat lost in joules (4.2 joules = 1 calorie), A is the surface area of the object and h is its thermal conductivity.[‡‡] Sticking all the numbers in and you end up with 135.2 joules or 0.032 kcal needed per second to maintain 416 °C. That isn't actually very much at all, certainly compared to the energy lost *via* radiating photons. A flaming Johnny is basically acting like an infrared heater, and that loses him 3227 kcal per second.[§§]

So in total, 10 minutes of flame time costs 29 183 kcal getting to temperature then another 1 936 277 kcal to maintain a toasty 416 °C. This means that, if Johnny Storm dropped in on Quicksilver for breakfast, the speedster would need to whip up another 3930 peanut butter sandwiches. Good job he can spread and slice at supersonic speeds!

1.6 SPIDER-MAN'S BREAKFAST

With all his leaping from building to building and wall climbing, Spider-Man will certainly need a nice high-calorie diet, but his energy intake is likely to be closer to an Olympic gymnast than a pyrokinetic or speedster. However, his silk-spinning power does have some very particular nutritional needs. Whether it is produced by fleas or flies, bees or beetles, crabs or spiders, silk is always made of protein. So, if Spider-Man spins copious amounts of web, then he must also consume

---

[‡‡]$\Delta T = 416 - 25 = 391$. $A = 1.9 \text{ m}^2$. Getting hold of the thermal conductivity of a super-heated human body is a bit tricky, so I hope you don't mind if I just use the value for hydrogen ($0.182 \text{ W m}^{-1} \text{ K}^{-1}$), after all if you count up the atoms 62% of a human is hydrogen.

[§§]In case you haven't had enough of equations by now then this is calculated by:

$$E = A \times \sigma \times (T)^4$$

where A is surface area, T is absolute temperature in kelvin and σ is something called the Stefan–Boltzmann constant and which has a value of $5.67 \times 10^{-8} \text{ W m}^{-2} \text{ K}^{-4}$.

huge amounts of protein.[¶] What, then, does Peter Parker have for breakfast to sustain his villain-beating lifestyle? To work this out, we need a bit of school-level physics, Maths and some basic biochemistry. First, we need to work out the strength of Spider-Man's silk, then calculate how much force Spider-Man would exert on the silk and, finally, figure out how much and what kind of protein will be needed to enable Spider-Man to do his stunts.

Let's assume Spider-Man produces threads that have the same characteristics as dragline silk produced by the European garden spider *Araneus diadematus*. That means it ought to have a tensile strength – which is the largest stress that a material can withstand before breaking – similar to that of a piano wire.[9] A piano wire has a tensile strength of 1.1 billion pascals, which is a unit of measure for pressure, or the force per unit area.

According to Marvel's directory,[10] Spider-Man weighs 75 kg. Applying Newton's second law, we can figure out the downward force that Spider-Man exerts, which is calculated by multiplying his mass by the Earth's gravitational acceleration (9.8 m s^{-2}). This works out to be about 735 newtons.

Now we just need to take the downward force exerted and divide it by the tensile strength of the silk. Without going into the tiny calculations,[ǁ] it turns out that Spider-Man could hang from the ceiling on silk that is less than 1 mm thick. (Of course, being a scientist and not an engineer, I've factored in exactly zero margin for error.)

Next, we need to establish what mass of silk fibre this would represent, assuming Spider-Man's pre-lunch heroics requires 100 m of it. Silk is slightly denser than water (1.3 g mL^{-1} compared to 1 g mL^{-1} for water), which means that 100 m of the silk would weigh about 87 g. There is about 6 g of protein in an egg. So, could it be that Spidey only needs about 15 eggs (87 g divided by 6) for breakfast if he plans to use 100 m of silk? That doesn't

---

[¶]Yes, I know that in most Spider-Man worlds Peter actually uses devices he wears on his wrists to spin his 'silk'. But there are versions of the Marvel Universe where he 'glands' the silk directly from his wrists.

[ǁ]Unless you really want to... The cross sectional area of spider silk required to support Spider-Man = $735 \text{ N}/1.1 \times 10^9 \text{ N m}^{-1} = 6.68 \times 10^{-7} \text{ m}^2 = 6.68 \times 10^{-3} \text{ cm}^2$. That works out to be a bit of silk with 0.046 cm radius, or just under 1 mm thick.

seem too bad. However, all proteins are not created equal, so the equation is not quite this simple.

All proteins are made from the same building blocks, amino acids. There are around 20 natural varieties of amino acids, with different sizes, shapes and chemical properties (Figure 1.2). These amino acids link together into chains, and the order and length of the chain is unique to a particular protein.

Spider silk is made, predominantly, of a protein called fibroin,[11] which is mostly made up of amino acids in the sequence glycine, serine, glycine, alanine, glycine, alanine repeated over and over again, forming long chains. These chains line up with each other and stick to their neighbours *via* masses of weak bonds, forming a structure known as a beta-sheet.

Other proteins are different. For instance, egg white is composed of a mixture of proteins, the main one being ovalbumin. Ovalbumin is made of 385 amino acids[12] and all 20 different amino acids feature multiple times.[13] Overall, it has a very different structure – composed of beta-sheets, plus corkscrew shapes called alpha-helices (Figure 1.3).

So, converting eggs to spider silk presents a bit of a problem, as the building blocks don't match up – their amino acid compositions are different. It's like having two model toy sets, a plane and a house, then trying to build the plane from the house kit. However, the body can do something with amino acids that can't be done with toy blocks. It has the ability to interconvert some amino acids. Another class of protein (with their own distinct shapes) called enzymes can whip bits off serine changing it to glycine or whack something onto serine and convert it to threonine (and *vice versa*, see Figure 1.4). We can take this interconversion of amino acids into account when figuring out how many eggs Spider-Man needs to make his silk.

In total, fibroin is 50% glycine, serine and threonine[14] whilst only 15% of the protein in hen eggs is made from these amino acids.[15] So, really, Spidey needs to consume over three times more egg protein than the silk protein he plans to use. This means he actually needs 50 eggs for his 100 m of silk. But that's not really the end of it either. After all, what happens if he leaps from a building to save a falling Mary-Jane and deploys his webslingers

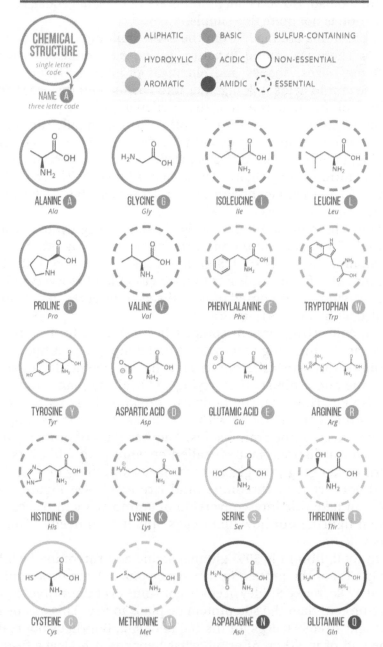

Figure 1.2 The 20 common naturally occurring amino acids come in a variety of flavours, some are acidic, others basic. A few have ring (aromatic) sidechains, several are hydrophobic (aliphatic). But they all share a common 'backbone' with a carboxylic *acid* and an *amino* group, hence the name 'amino acid'. © Andy Brunning 2017.

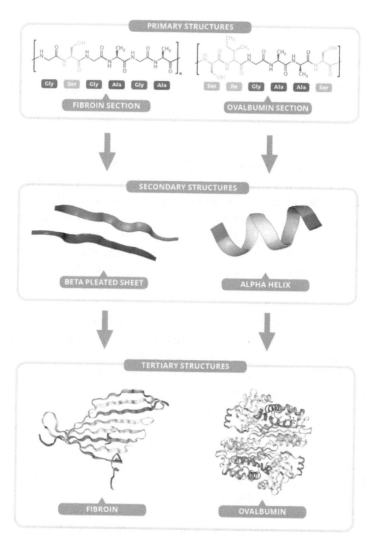

Figure 1.3　Amino acids are linked together *via* their amino and acid groups to form long chains. The order of the amino acids is known as the primary structure. These chains then fold up into sheets and helices, which are known as the secondary structures. These in turn come together to make the final tertiary three-dimensional structure of a protein. In these images all the individual atoms have been removed so that the path of the chains can be more clearly seen. © Andy Brunning 2017.

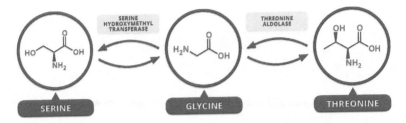

Figure 1.4   Enzyme conversion of amino acids. © Andy Brunning 2017.

to save the day*** (as he does about halfway through the 2002 movie with Tobey Maguire as Spider-Man). Now, I'm guessing he's still using his drag silk (which stretches by 27%). He could go for the flag silk; it's not as strong (tensile strength = 0.5 GPa) but it's got amazing elastic properties as it can stretch to 2.7 times its original length![9] But that might be a bit too bouncy.

In the scene, Spidey leaps from the balcony and falls for seven seconds before his silk starts to arrest his fall. He has caught M.-J so let's say their combined weight is 125 kg. How much silk is he going to need here?

From the time of Spidey's fall, we can calculate that he fell about 240 m (wow, that's one high balcony). Plus, assuming the silk stretches to its maximum, it will be a fall of about the height of the Eiffel Tower. The impact force on the silk rope as they slow down would be about four times their mass.††† This force is about six times greater than the force when Spidey is just hanging around on the end of his line of silk. However, taking into account the length of the fall (240 m) and the extra force, he will need 1.3 kg of silk to catch his fall. So, he must have had about 750 eggs for breakfast that morning, just to have enough silk for that one scene. I think Aunt May might have noticed!

Of course, there is an alternative. Being the genius that he is, Peter Parker should be fully aware of how spiders behave. Silk-spinning

---

***Overlooking, of course the fact that gravity works on Spider-Man, accelerating him towards the ground at the same rate as M.-J. So with her 1 second head start she's already fallen 5 metres, a distance that physics says Spidey can't make up. Or maybe he has some sort of mini-rocket boosters in his boots that we don't know about?

†††Spidey accelerates at 9.8 m s^{-2} for 7 seconds, giving him a velocity of 68.6 m s^{-1}. Given that velocity $= \sqrt{(2 \times g \times h)}$, we can work out Spidey fell for 240 m. With silk stretching to its maximum (27%) that gives us a stopping distance of 64.8 m. The impact force on the silk rope as they slow down is $F = 1/2mv^2/d = 4538 \text{ N}$ (where d is the stopping distance, m is mass and v is velocity).

arachnids don't waste their precious silk; they recycle it, eating up old broken webs so that it can be digested and re-spun. Perhaps there are scenes on the cutting room floor, showing Spider-Man tidying up the disaster areas in which he was embroiled and slurping up his used web like so much spaghetti, only to be used again. If so, this could change all of our equations considerably!

REFERENCES

1. Food and Drink Federation, Front of pack labels, Using the front of pack label, *Guideline Daily Amounts*, Foodlabel.org. uk, 2016 [cited 30 October 2016], available from: http://www. foodlabel.org.uk/label/gda_values.aspx.
2. M. Looi, *How Olympic Swimmers Can Keep Eating Such Insane Quantities of Food*, [Internet], Quartz, 2016 [cited 30 October 2016], available from: http://qz.com/753956/how-olympic-swimmers-can-keep-eating-such-insane-quantities-of-food/.
3. R. Allain, *Who's Faster? Flash or Quicksilver?* [Internet], WIRED, 2016 [cited 30 October 2016], available from: https:// www.wired.com/2014/06/whos-faster-flash-or-quicksilver/.
4. W. McArdle, F. Katch and V. Katch, *Exercise Physiology*, Lippincott Williams and Wilkins, Philadelphia, 8th edn, 2014.
5. Pietro Maximoff (Earth-616), [Internet], *Marvel Database*, 2016 [cited 30 October 2016], available from: http://marvel. wikia.com/wiki/Pietro_Maximoff_(Earth-616).
6. Jonathan Storm (Earth-616), [Internet], *Marvel Database*, 2016 [cited 13 October 2016], available from: http://marvel. wikia.com/wiki/Jonathan_Storm_(Earth-616).
7. J. Sallabank, C. Sullivan, A. Foden and A. Higgins, *J. Phys. Spec. Top.*, 2014, **13**(1), A4_3, [cited 13 December 2016], available from https://physics.le.ac.uk/journals/index.php/pst/article/view/728.
8. I. Newton, *Isaaci Newtoni Opera quae exstant omnia*, Excudebat Joannes Nichols, 1st edn, 1782, vol. 4.
9. L. Römer and T. Scheibel, The elaborate structure of spider silk, *Prion*, 2008, **2**(4), 154–161.
10. Spider-Man (Peter Parker)–Marvel Universe Wiki, *The Definitive Online Source for Marvel Superhero Bios*, [Internet], Marvel.com, 2016 [cited 30 October 2016], available from: http://marvel.com/universe/Spider-Man_(Peter_Parker).

11. M. Levy and E. Slobodian, Sequences of amino acid residues in silk fibroin, *J. Biol. Chem.*, 1952, **199**(2), 563–572.
12. A. Nisbet, R. Saundry, A. Moira, L. Fothergill and J. Fothergill, The complete amino-acid sequence of hen ovalbumin, *Eur. J. Biochem.*, 1981, **115**(2), 335–345.
13. J. E. Hicks, *A Study of the Amino Acids Associated with Ovalbumin*, Master of Science Thesis, Texas Tech University, 1970, available from: https://ttu-ir.tdl.org/ttu-ir/bitstream/handle/2346/14771/31295010774445.pdf?sequence=1.
14. S. O. Andersen, Amino acid composition of spider silks, *Comp. Biochem. Physiol.*, 1970, **35**(3), 705–711.
15. M. Minnaar, *Green Eggs, Anyone? Emu's Zine*, available from: http://www.emuszine.com/Food/green_eggs_anyone_by_maria_minnaar.htm.

The Evolution of Superpowers

LOUISE K. GENTLE

School of Animal Rural and Environmental Sciences, Nottingham Trent University, UK
E-mail: louise.gentle@ntu.ac.uk

"... from so simple a beginning, endless forms most beautiful and most wonderful have been, and are being, evolved."[1]

2.1 INTRODUCTION

Superheroes seem to have quite similar background stories – either they were born with their superpowers (such as the X-Men), or they were involved in some sort of radiation accident (the Hulk, Spider-Man, *etc.*) which caused them to mutate or evolve rapidly, resulting in the acquisition of superpowers. However, in either case, we might ask what were the circumstances that caused the evolution of these superpowers? Superpowers, along with many other traits, can be viewed as beneficial to the individuals who possess them, even if many superheroes struggle to see the benefit of their powers. So, why and how do new traits such as superpowers evolve? Well, evolutionary theory dictates that if

The Secret Science of Superheroes
Edited by Mark Lorch and Andy Miah
© The Royal Society of Chemistry, 2017
Published by the Royal Society of Chemistry, www.rsc.org

an organism evolves a feature that is environmentally advantageous – in that it increases the chances of reproduction – then
this advantage will be passed on to the next generation, but the
manner in which that takes place varies.

2.2   NATURAL SELECTION

Evolution can be a change in morphology (*e.g.* shape), ecology
(*e.g.* where it can survive), behaviour (*e.g.* how it attracts members
of the opposite sex), or physiology (*e.g.* what food it can tolerate).
Darwin's theory of natural selection rests on three observations.

1. There is a lot of variation within individuals of a population. If we take human height as an example, there are tall
   people and short people and a good spread of individuals in
   between. There are very few individuals who are extremely
   tall or extremely short, but these individuals do exist.
2. Parents produce more offspring than are needed to replace
   themselves, so populations tend to increase.
3. Environments are limiting, only allowing enough food,
   shelter, and breeding space to support a population of a
   particular size. This creates competition, so members of a
   population must compete to survive. Within any population, some individuals will be better suited to an environment than others. So, the best-adapted individuals survive
   and reproduce. If the advantageous characteristics are heri-
   table, then they can be passed to the next generation – the
   best-adapted individuals increase at the expense of the
   least-adapted individuals.

So, natural selection can explain both evolution and adaptation, a good example of which is found in the giraffe. If all
giraffes had necks of the same length, then they would all only
be able to reach up to a certain point in the trees. However, if
just one individual has a slightly longer neck it would have a
competitive advantage – it would be able to reach higher in the
trees than all other giraffes, so enabling it to access the leaves
that other giraffes couldn't. In this way, the tallest giraffe would
be able to acquire more food than the other giraffes – enough
food to enable its survival. Survival gives it more time in which

to reproduce. Furthermore, if the longer neck is something that is inherited, the offspring of the long-necked giraffe would also have longer necks and a consequent competitive advantage over other giraffes in the population. The giraffe with the longer neck is selected for naturally. So, organisms evolve and adapt over time to make them function efficiently in their environment.

Evolution is generally a slow business and the time we have had to actually observe evolution is very short. However, there are cases where a natural population has changed its form over a short enough period of time for natural selection to be observed. Darwin's finches are one such example. During 1977, there was a severe drought across the Galápagos Islands which caused a rapid decline in the number of small seeds available for Darwin's medium ground finch (*Geospiza fortis*) to feed upon. This meant that only the larger and harder shelled seeds were present in the environment, which the majority of finches couldn't crack. Consequently, the medium ground finch declined by a massive 85% that year. The remaining 15% of birds were considerably larger than those that died, and had particularly large beaks – perfectly adapted to crack their way into the larger seeds.[2] Consequently, only these individual birds procreated, leading to a resulting impact on all subsequent generations.

2.3 GENETICS

So, we now know *why* traits evolve and the evolutionary pressures causing them, but *how* would unique traits or superpowers actually evolve? Apart from identical twins or clones, individuals have unique DNA (deoxyribonucleic acid), which carries the genetic code that determines an organism's characteristics, rather like a recipe for a cake. The DNA is arranged into *chromosomes* and, within each chromosome, regions called *genes*. Each gene contains the information necessary to make a particular protein, and it is the proteins that are the units that determine an organism's characteristics, from the anatomy of the stomach to the colour of its hair.

Most of a body's cells contain two sets of chromosomes, one inherited from the mother and one from the father. Some genes have different versions, which we call *alleles*, of which there may be many versions. For example, there are several different alleles for eye colour, giving a variety of colours from green to blue to brown. Nevertheless, to help explain this process further, imagine that there are just two alleles for eye colour: brown and blue. An individual will inherit two alleles, which could be two brown, or two blue, or one brown plus one blue. Sometimes, alleles can be co-dominant where the effect of them is merged, but sometimes one can dominate or mask the other. Eye colour is often used as an example of *dominance* (although in truth the genetics of eye colour is much more complicated than this), with the brown allele dominating the blue allele; the blue allele is then said to be *recessive*. So, if you have two brown alleles you would have brown eyes, if you have two blue alleles you would have blue eyes, but if you have one brown allele and one blue allele you would have brown eyes as the brown allele dominates the blue allele. Of course, genetics isn't always that easy, but you get the point here (see Figure 2.1). In this case, it is only when you get *two* copies of the recessive allele that they are actually expressed (or seen). So, two brown-eyed parents can produce a blue-eyed

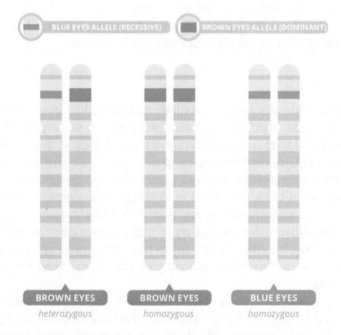

Figure 2.1   A simplified view of the inheritance of eye colour. © Andy Brunning 2017.

child (but only if both parents have one brown allele and one blue allele each). In this way, two parents could create a 'mutant' and, if you had two 'mutant' parents, you would produce 'mutant' offspring, as occurs in *The Incredibles*, where all members of the family possess superpowers.

So, we know that individuals have unique genetic information but how does this actually arise? Firstly, there is independent assortment of chromosomes. Your gametes (sex cells: sperm or eggs) contain only half of your genetic material. This half of your genetic material is assigned randomly, regardless of whether you inherited it from your mother or father, and ensures that all your gametes are different. This is why your siblings are not identical to you and why your offspring aren't identical to each other. Secondly, a process called *recombination* can occur when gametes are produced. This is where the pairs of chromosomes become stuck together and essentially exchange DNA when they are pulled apart to produce the gametes. So, instead of producing a gamete that contains a chromosome from just your father, you might end up with a chromosome which also contains a little bit of genetic material from your mother – so you produce a gamete that has a mixed up, but novel, genetic content. The final way to produce individuals with unique DNA is *via* mutations. A mutation is a mistake made during DNA replication – unfortunately, nothing is perfect, including our own bodies, so mistakes occur. In fact, mistakes are so common that there may be 100–200 new mutations every time DNA is passed from one generation to the next.[3] These mutations may have effects on the individual, but they create distinctive *genotypes* (the genetic makeup of an organism). Mutations can also occur as a result of exposure to mutagens such as radiation. This is where most of the superheroes acquired their powers – Peter Parker was bitten by a radioactive spider at a science exhibit, which gave him the agility and proportionate strength of a spider (Spider-Man); Bruce Banner was exposed to gamma radiation during the detonation of an experimental bomb, which gives him seemingly limitless strength when he morphs into the Hulk when subjected to stress; the Teenage Mutant Ninja Turtles were covered in radioactive ooze, which allowed rapid evolution and mutation into more human forms, and the Fantastic Four were exposed to cosmic radiation during the test flight of an experimental spacecraft.

2.4   NON-SELECTIVE EVOLUTION

It is the continued generation of novel genotypes that allows evolutionary processes such as selection to operate. The genotype is the set of genes possessed by an individual (*e.g.* one brown eye colour allele plus one blue eye colour allele) and the *phenotype* is the outward appearance of an individual (in this case, brown eyes). Evolution is a change of genotype through time, but the phenotype doesn't necessarily change. In any case, what we have discussed so far is a form of evolution that is very familiar. This is active evolution by selection, *i.e.* where a change in the environment results in pressures that end up selecting for a particular characteristic. In the case of the medium ground finch, the drought was the environmental change and it resulted in birds with larger beaks surviving.

However, there are other ways in which the genotype can evolve in the absence of selective pressure – evolution can also happen in the background. There are three main mechanisms by which this can occur.

1. *Gene flow*, where changes occur in the gene pool (the complete set of genes) of a population, usually because of the introduction of genes from another population by immigration. For example, if two lion prides met and ended up interbreeding, their gene pools would both change. Gene flow can even happen between related species: recently, polar-grizzly bear hybrids (which have white fur, characteristic of polar bears, but long claws and brown facial patches, characteristic of grizzly bears) have been identified in the wild[4] and one possible explanation for this is that the polar bears have been driven southwards due to global warming and the melting of the polar ice caps.

2. *Random genetic drift*, which describes random movement in gene frequencies between populations. For example, if 50% of a population are red and 50% are green you would expect that the next generation will also be in the ratio 50:50. However, due to random sampling the next generation could be 49:51. In extreme cases, particularly in small populations, certain genes may either be lost or expressed through genetic drift.

3. *Founder effects* occur when just a few original founders establish a new population. These will carry only a small fraction of the genetic variation of the original population, and means that the frequency of rare recessive genes can be greatly enhanced. For example, the red kite (*Milvus milvus*) population in Wales was reduced to just a handful of individuals, resulting in a highly inbred population that experienced reproductive difficulties until conservation measures were put in place.[5] Similarly, the Florida panther (*Puma concolor coryi*) is a small population of puma that has been isolated from other puma populations through habitat destruction and fragmentation. Consequently, it has very low genetic diversity and has a morphological anomaly, a kinked tail, which is rarely seen in any other puma species.[6] The presence of a kinked tail doesn't give the Florida panther any sort of superpower, but the evolution of rare phenotypes, such as superpowers, are more likely to appear in small isolated populations.

All of this brings us to the fact that change is necessary, not to increase adaptiveness, but to maintain it.[7] This has been termed the Red Queen Hypothesis and is derived from a statement that the Red Queen made to Alice in Lewis Carroll's *Through the Looking-Glass*:[8]

> ... the Red Queen and Alice were running hand in hand, and the Queen went so fast that it was all Alice could do to keep up with her... however fast they went, they never seemed to pass anything...just as Alice was getting quite exhausted they stopped... Alice looked around her in great surprise. 'Why, I do believe we've been under this tree the whole time! Everything's just as it was!' 'Of course it is' said the Queen... 'here you see, it takes all the running you can do to keep in the same place'.

So, evolution is a gradual change – there may be no physical change in phenotype whilst the underlying genotype is changing gradually. So how would mutants suddenly appear in a population? Well, they could be exposed to some sort of radiation, or they could have been subjected to some founder effects, increasing the expression of rare recessive genes. But are there any other ways?

Fossil records indicate that there may be other processes that occur. Although the chance of an individual becoming fossilized is very low, and the chance of finding a fossil showing a gradual evolutionary process is consequently even lower, these transitional changes are seen. For example, the discovery of Archaeopteryx was, at the time, the missing link between reptiles (specifically dinosaurs) and birds. Archaeopteryx is an organism that possesses features including a toothed jaw (from dinosaurs) and flight feathers (from birds).

Nevertheless, once some species appear in the fossil record, they show little evolutionary change, appearing relatively stable over the vastness of geological time. So, species seem to suddenly appear in the fossil record and remain for millions of years, then suddenly disappear with no change to their appearance – this is termed *punctuated equilibrium*.[9] A current example of this could be the coelacanths (*Latimeria* sp.) which are closely related to lungfish and are thought to be an evolutionary transitional species between fish and tetrapods (four-legged creatures). Coelacanths were thought to have gone extinct at the same time as the dinosaurs (65 million years ago), but in 1938 one was discovered off the coast of South Africa.[10]

Species with large populations are generally stable because new mutations are diluted by the population's large size and a constantly fluctuating environment. However, mutations in smaller populations are much more likely to show. This is because they are usually isolated populations under additional pressure from natural selection, as they exist at the outer edges of ecological tolerance. If most evolution happens in these small, isolated populations, evidence of gradual evolution in the fossil record should be rare. The evolution of traits that are superpowers are more likely to appear in these small, isolated populations. So, perhaps the X-Men mutants, who were born with superhuman abilities, originated from some sort of founder population, where rare alleles were more likely to be seen.

An alternative way for mutants to suddenly appear in a population would be due to something called epigenetics, where variations in the environment can cause changes in the organism's phenotype. For example, turtles and crocodiles undergo temperature-dependent sex determination, where the temperature of the environment determines the sex of their undeveloped eggs – in turtles, eggs from cooler nests hatch as males, whereas eggs from warmer nests are female.[11] So, perhaps superheroes developed in a specific environment that somehow activated their dormant superpowers, but how could this work in practice?

Aquaman possesses a number of superpowers, including the ability to breathe underwater, but in what scenario might this ability evolve? Life evolved in the seas where many animals use gills to take in oxygen, so it is no surprise that embryos go through a stage where they have slits or openings in the neck that resemble gills. These ancestral characters do not develop into gills – they are only present in the embryo for a short while. However, hypothetically, if there was a certain environmental condition, perhaps a particularly high temperature, which caused the gills to develop and led to the individuals being able to breathe underwater, then the evolution of one of Aquaman's superpowers would be achieved.

2.5 SUPERPOWERS

In sum, evolution has been occurring for billions of years and has produced organisms that are perfectly adapted to their environments. It has produced crawlers, wrigglers, jumpers, walkers, runners, fliers, gliders, and these are just some of the variations in movement. But, has it already produced some superpowers? Of course it has!

2.5.1 Vision

Humans are able to see due to the presence of photoreceptors – cells that convert light into signals. These come in two basic types: rods and cones. The human retina contains around 120 million rod cells and 6 million cone cells.[12] In contrast, nocturnal species such as owls have an even higher density of rods in their retinae, to help them see in the dark. The superhero Doctor Midnight has the ability to see in darkness at the cost of near or total blindness in sunlight – so, like owls, he must have a high density of rods in his retina.

Many diurnal birds of prey (those that are active during daylight) also have much better visual acuity than humans. To start with, they have very large eyes in relation to the size of their head and they have tube-shaped eyes that produce a larger retinal image. Also, like owls, they have a much higher density of photo-receptors in the retina – the more receptors an animal has, the higher its ability to distinguish individual objects at a distance. For example, the American kestrel (*Falco sparverius*) has 65 000 photoreceptors/mm^2 compared to 38 000/mm^2 in humans, so it is theorized that it can see an insect of just 2 mm in size from the top of an 18 m tree.[13] This keen 'telescopic eyesight' is also adopted by X-Men's Peepers, a member of Magneto's brotherhood of evil mutants who is reported to identify objects way beyond the normal range of vision, and Nova, allied to the Fantastic Four, who can apparently locate specific objects in the vastness of space!

Additionally, humans have three types of cone photoreceptors, those sensitive to short wavelengths around 420 nm (typically the colour blue), medium wavelengths of 534 nm (associated with the colour green) and long wavelengths of 563 nm (the colour red), to give us tri-chromatic vision and the range of colours that we are able to detect – all of which are variations of blue, green and red. This 'human-visible' part of the electromagnetic spectrum is 400–700 nm. However, birds commonly have four or even five types of cones, allowing their vision to be far superior than ours. Thus, most birds can see in ultraviolet (UV) wavelengths (300–400 nm) to which humans are 'blind'. This UV vision has been found to be used for orientation and navigation, foraging or the selection of potential mates whose plumage can look even more striking when viewed under UV light. For example, male and female blue tits (*Cyanistes caeruleus*) have very similar plumage to our eyes but look completely different to each other when viewed under UV light.[14] Amazingly, the common kestrel (*Falco tinnunculus*) can detect UV reflections of vole urine and use this cue to confine hunting behaviour to areas with high densities of prey.[15] So, the kestrel sees the ultraviolet urine trail path of a rodent and knows where to hunt. Could a superhero also use this power to his or her advantage? Well, it is alluded to that several superheroes are able to see into the UV spectrum, including Superman, Photon (one of the Avengers) and Doctor Manhattan.

As well as being able to see into the UV part of the electromag-netic spectrum, there are a number of non-humans, notably reptiles, that can detect much longer wavelengths in the infrared spectrum and thus 'see' a thermal image.[16] Vipers, pythons and boas possess pit organs that contain a membrane that senses heat from the warm bodies of prey items a short distance away – particularly useful for finding food at night. This nifty superpower is also shared by Supergirl, Wonder Woman and Lar Gand (one of Superman's allies). Indeed, supervillain Doctor Doom should really look into acquiring this power, so that he can sense the Fantastic Four's Human Torch coming!

2.5.2 Hearing

As nocturnal animals, bats cannot see in the dark, so they use hearing to navigate and locate prey *via* a process termed echolocation. This is where the bats emit high-frequency sounds (around 50 kHz, much higher than human hearing which is up to 20 kHz) and listen for the echoes that bounce off objects. The difference in time between emitting the sound and hearing the sound bounce back allows the bats to build up a 'picture' in their brains of their surrounding environment – sounds that take longer to bounce back indicate that the surroundings are further away. Once a bat locates a prey item it emits sounds at a faster rate, allowing it to pinpoint its prey with a high degree of accuracy until it is close enough to grab it. It isn't just bats that use echolocation, other species such as toothed whales, shrews and swiftlets (cave-dwelling birds) use a cruder version of echolocation – probably more to investigate their habitats and navigate rather than to locate prey.

This crude version of echolocation is a sense that has also developed in some humans, particularly those who are blind and so use acoustic cues as a way of navigating their environments. These super humans create sounds, such as making clicking noises with their mouths, or tapping canes on the ground, then listen for the echoes when the sounds get reflected off objects, enabling them to identify how far they are from various items. So, in the superhero world, it is not surprising that Matt Murdock, who was blinded by radioactive waste as a child, developed an echolocation sense and became the superhero Daredevil. Although Daredevil is blind, his other senses have become heightened as they compensate for the lack of vision.

In fact, the echolocation sense gives him an omnidirectional field of vision, allowing him to locate objects or people in all directions – an obvious advantage over normal human vision.

2.5.3   Electric and Magnetic Senses

Species such as the electric fish are capable of producing electric fields from a special organ, a structure of modified nerve cells that produce stronger electric fields than usual. The discharge from this electric organ varies. For example, weakly electric fish such as the elephantnose fish (*Gnathonemus petersii*), which lives in the dark, murky rivers of Africa, generate an electric discharge that is typically less than one volt in amplitude.[17] These discharges are used in a similar way to the echolocation sense of bats, to navigate in an environment where vision is redundant, detect objects and communicate with other electric fish. The brown ghost fish (*Apteronotus leptorhynchus*) makes different signals depending on whether they are interacting with a fish of the same or opposite sex, so electrocommunication may be involved in courtship rituals![18] However, strongly electric fish, such as the electric eel (*Electrophorus electricus*), have electric organ discharges of several hundred volts – so powerful that they can use them to stun their prey. This is not quite as good as the power of supervillain Electro (one of Spider-Man's enemies), who can control electricity and emit millions of volts per discharge – enough to knock his foes dead!

The contraction of muscles is associated with an electrical discharge, which means that most living organisms produce an electric field, albeit small. Nevertheless, many species that are not able to produce larger amounts of electricity themselves are still capable of detecting this small electric field. Most notable are cartilaginous fish, such as sharks and rays, who have evolved specialized receptors called ampullae of Lorenzini to detect the muscle contractions of their prey – particularly useful for detecting animals that bury themselves in sand. However, the detection takes place only over very short ranges, possibly a few centimetres. So, perhaps Spider-Man could hunt for his enemy Sandman in the same way that sharks and rays scan the sand for fish? The sensory organs of sharks also allow them to detect magnetic fields such as that of the earth as the presence of the earth's magnetic field induces an electric field in the ocean currents. Thus, sharks and rays can orient to the electric fields of oceanic currents and may be able to sense their magnetic heading.[19]

A recent shark defence study involved attracting free-swimming wild southern stingrays (*Dasyatis americana*) and nurse sharks (*Ginglymostoma cirratum*) to a feeding platform, one side of which was magnetized, and found that the rays and sharks avoided the magnetized side of the apparatus.[20] Perhaps the reluctance of sharks to go near magnets is something that Batman has known for a while – presumably magnetic particles are the main ingredient in his famous shark-repellent spray!

However, sharks aren't the only animals capable of detecting magnetic fields. Magnetoreception is used by various species throughout the animal kingdom to assist navigation and orientation, a great example of which are homing pigeons (*Columba livia domestica*). These amazing birds are capable of navigating back to their home lofts in many different conditions from very sunny to completely overcast weather. The presence of magnets disrupts this behaviour, suggesting scientifically that they may be using the earth's magnetic field to navigate.[21] Although the exact processes are not completely understood, homing pigeons have been found to possess a substance called magnetite in their beaks which becomes magnetized when exposed to magnetic fields.[22] Magnetoreception seems to be a fairly common power in the superhero world too, with several superheroes and villains capable of sensing and manipulating magnetic fields. Most prominent are the X-Men villain Magneto and his daughter Polaris. Whilst Magneto's control of magnetism can produce a wide range of effects from lifting and manipulating metal objects to rearranging matter, Polaris' abilities also include perceiving the world as patterns of magnetic and electrical energy and being able to detect living organisms – not too dissimilar to the powers already present in the animal kingdom.

2.5.4 Shapeshifting

The ability to change shape and mimic others – shapeshifting – is a formidable power for a superhero or villain and has been used by X-Men's Mystique on many occasions to lure and manipulate her foes. Although animals come in all different shapes and sizes, only a few are capable of changing their shape. When an animal is under threat from a predator, it can fight, run or hide. Sometimes, running might actually make the animal more obvious to its attacker. However, an animal that happens to look the same as its environment may survive by being camouflaged from the attacker and we see this occur in moths that resemble fallen leaves, or chameleons that blend in with their background.

Many animals have evolved such permanent adaptations that help them to mimic their environments. Nevertheless, there are a few animals known to be capable of changing their shape. For example, the mimic octopus (*Thaumoctopus mimicus*) was discovered in 1998 off the coast of an Indonesian island, and is perhaps the greatest shapeshifter of all.[23] Similar to the cuttlefish, it is capable of mimicking its background environment by changing the colour and texture of its skin. However, it is the only animal able to mimic a diverse range of species – at least 13 have been recorded so far – including lion fish, sea snakes, jellyfish and sea anemones.[24] The mimic octopus has remarkable dexterity, being capable of changing its colour, behaviour, shape and texture, and can alter its mimicry according to the circumstances. Most of the impersonated species are poisonous, giving the mimic octopus protection from predators, but it is also able to imitate prey items, possibly luring them in before feasting on them.

2.5.5 Absorbing Powers

The ability to steal the superpowers of another individual is surely the ultimate power, enabling one to enjoy any of the superpowers in existence! X-Men's Rogue has the ability to absorb the superpowers of anyone she touches, and this too has some biological precedent in the way that some animals become toxic due to the food that they eat. For example, the larvae of the cinnabar moth (*Tyria jacobaeae*) feed on ragwort (*Senecio jacobaea*), a common wild plant known to be poisonous to animals such as horses and cattle.[25] However, the toxins present in ragwort have no detrimental effect on the moth larvae. In fact, the larvae are able to absorb and sequester the toxins,[26] making themselves noxious, and hence unpalatable to predators such as insectivorous birds. This accumulation of 'powers' may be one way in which superheroes could undertake power absorption, but ingesting another individual isn't something that superheroes such as Rogue do, so we need to look elsewhere for a biological principle that could explain this. The pitohui birds of New Guinea, such as the hooded pitohui (*Pitohui dichrous*), are possibly the only poisonous birds in the world. Their feathers and skin contain a noxious batrachotoxin substance, similar to that secreted by the poison dart frogs, which gives them defence against predators. It is thought that, like the cinnabar moth, the birds cannot produce the toxic substance directly – instead, they acquire it *via* Choresine beetles that they feed upon.[27] In addition, it is suggested that this toxicity might be able to rub off onto the eggs or young of the pitohui birds, so the eggs and young in the nest are also toxic to predators, even though they have yet to feast on the Choresine beetles.[28] Thus, the superpower toxicity of the birds is so great that it can be transferred between individuals – the young are essentially absorbing the superpower from their parents in the same way that Rogue absorbs powers when she touches other people.

2.5.6 Chemical Weapons

A further example of a natural superpower is found in the bombardier beetles, namely the bulls-eye beetle (*Stenaptinus insignis*). These beetles are noted for their unique defence mechanisms whereby they produce small acid gas bombs to deter their predators such as ants. In extreme cases they may even bombard their predators with so much of the chemical bombs that they kill them.[29] The acid bombs consist of two chemical compounds, hydroquinone and hydrogen peroxide, that are stored separately in the beetle's abdomen. When threatened, the beetles are able to combine the two compounds, resulting in the production of a boiling mixture that explodes out of the tip of the directional abdomen as a gas, accompanied by a popping or crackling sound. This remarkable weapon is surely the precursor for a superpower! The superheroes Anarchist and Zeitgeist (members of X-Force) both have mutant acid generation powers. Anarchist secretes an acid-like sweat and, in enough quantities, can produce corrosive blasts of acid that he shoots from his hands, whereas Zeitgeist simply spews acidic vomit! Perhaps it is time that some of the superheroes caught up with the amazing set of powers that have already evolved in the animal kingdom and produced acid bombs.

2.6 CONCLUSIONS

There is lots of individual variation between people – some are tall, some are short; some have blue eyes, some have brown eyes; some can hear bats or ultrasonic pet deterrents, whilst others cannot. Most populations follow what is called a normal distribution where the majority of people are average for a trait, but there are a few outliers, for example extremely tall or short people. The hearing range of humans follows a similar pattern. So, superpowers may just be extreme variation within individuals – there are humans with extraordinary senses capable of echolocation or even detecting certain diseases.[30] In many cases, a superpower found in comic books or in films has a natural precedent in some other species, and being superhuman may simply require having one of these other natural capacities enjoyed by other animals, rather than having to invent something completely unknown. However, there are also humans with milder superpowers that are worth considering too. My own superpower is that I can smell sugar in tea, something that amuses a lot of my students. I can't imagine how I would ever use it to save the universe but I hope that I will use my powers for good rather than evil!

REFERENCES

1. C. Darwin, *On the Origin of Species by Means of Natural Selection, or the Preservation of Favoured Races in the Struggle for Life*, John Murray, London, 1859.
2. P. T. Boag and P. R. Grant, Intense natural selection in a population of Darwin's finches (*Geospizinae*) in the Galápagos, *Science*, 1981, **214**(4516), 82–85.
3. Y. Xue, Q. Wang, Q. Long, B. L. Ng, H. Swerdlow, J. Burton, C. Skuce, R. Taylor, Z. Abdellah, Y. Zhao, Asan, D. G. MacArthur, M. A. Quail, N. P. Carter, H. Yang and C. Tyler-Smith, Human Y chromosome base-substitution mutation rate measured by direct sequencing in a deep-rooting pedigree, *Curr. Biol.*, 2009, **19**(17), 1453–1457.
4. J. Mallet, Hybridization, ecological races and the nature of species: empirical evidence for the ease of speciation, *Philos. Trans. R. Soc. London, Ser. B*, 2008, **363**(1506), 2971–2986.
5. P. Davis, The red kite in Wales: setting the record straight, *Br. Birds*, 1993, **86**(7), 295–298.

6. B. Keim, Texan tail saves Florida panthers, for now, *Wired*, 2010, available at https://www.wired.com/2010/09/florida-panthers/. Accessed 12th June 2017.

7. L. Van Valen, A new evolutionary law, *Evol. Theory*, 1973, **1**, 1–30.

8. L. Carroll, *Through the Looking-Glass*, Macmillan, London, 1871.

9. N. Eldredge and S. J. Gould, Punctuated equilibria: an alternative to phyletic gradualism, in *Models in Paleobiology*, ed. T. J. M. Schopf, Freeman Cooper, San Francisco, 1972, pp. 82–115.

10. J. L. B. Smith, *Old Fourlegs: The Story of the Coelacanth*, Longmans, Green and Co., London, 1956.

11. A. Quinn, How is the gender of some reptiles determined by temperature? *Sci. Am.*, 2007, available at https://www.scientificamerican.com/article/experts-temperature-sex-determination-reptiles/. Accessed 12th June 2017.

12. G. Osterberg, Topography of the layer of rods and cones in the human retina, *Acta Ophthalmol.*, 1935, **13**(6), 1–102.

13. G. C. Whittow, *Sturkie's Avian Physiology, 5th Edition*, Academic Press, London, 1998.

14. S. Hunt, A. T. D. Bennett, I. C. Cuthill and R. Griffiths, Blue tits are ultraviolet tits, *Proc. R. Soc. B*, 1998, **265**, 451–455.

15. J. Viitala, E. Korpimäki, P. Palokangas and M. Koivula, Attraction of kestrels to vole scent marks visible in ultraviolet light, *Nature*, 1995, **373**, 425–427.

16. E. O. Gracheva, N. T. Ingolia, Y. M. Kelly, J. F. Cordero-Morales, G. Hollopeter, A. T. Chesler, E. E. Sánchez, J. C. Perez, J. S. Weissman and D. Julius, Molecular basis of infrared detection by snakes, *Nature*, 2010, **464**, 1006–1011.

17. P. Moller, *Electric Fishes: History and Behavior*, Chapman & Hall, London, 1995.

18. H. Zakon, J. Oestreich, S. Tallarovic and F. Triefenbach, EOD modulations of brown ghost electric fish: JARs, chirps, rises, and dips, *J. Physiol. (Paris)*, 2002, **96**, 451–458.

19. A. P. Klimley, S. C. Beavers, T. H. Curtis and S. J. Jorgensen, Movements and swimming behavior of three species of sharks in La Jolla Canyon, California, *Environ. Biol. Fishes*, 2002, **63**, 117–135.

20. C. P. O'Connell, D. C. Abel, P. H. Rice, E. M. Stroud and N. C. Simuro, Responses of the southern stingray (*Dasyatis americana*) and the nurse shark (*Ginglymostoma cirratum*) to permanent magnets, *Mar. Freshwater Behav. Physiol.*, 2010, **43**(1), 63–73.

21. W. T. Keeton, Magnets interfere with pigeon homing, *Proc. Natl. Acad. Sci.*, 1971, **68**(1), 102–106.

22. C. V. Mora, M. Davison, J. M. Wild and M. M. Walker, Magnetoreception and its trigeminal mediation in the homing pigeon, *Nature*, 2004, **432**(7016), 508–511.

23. M. D. Norman and F. G. Hochberg, The mimic octopus (*Thaumoctopus mimicus* n. gen. et sp.), a new octopus from the Tropical Indo-West Pacific (Cephalopoda: Octopodidae), *Molluscan Res.*, 2005, **25**, 57–70.

24. M. D. Norman, J. Finn and T. Trehenza, Dynamic mimicry in an Indo-Malayan octopus, *Proc. R. Soc. London, Ser. B*, 2001, **268**, 1755–1758.

25. A. M. Craig, E. G. Pearson, C. Meyer and J. A. Schmitz, Clinicopathologic studies of tansy ragwort toxicosis in ponies: sequential serum and histopathological changes, *J. Equine Vet. Sci.*, 1991, **11**(5), 261–271.

26. S. E. W. Opitz and C. Müller, Plant chemistry and insect sequestration, *Chemoecology*, 2009, **19**, 117.

27. J. P. Dumbacher, A. Wako, S. R. Derrickson, A. Samuelson, T. F. Spande and J. W. Daly, Melyrid beetles (Choresine): a putative source for the batrachotoxin alkaloids found in poison-dart frogs and toxic passerine birds, *Proc. Natl. Acad. Sci.*, 2004, **101**(45), 15857–15860.

28. J. P. Dumbacher, T. F. Spande and J. W. Daly, Batrachotoxin alkaloids from passerine birds: a second toxic bird genus (*Ifrita kowaldi*) from New Guinea, *Proc. Natl. Acad. Sci.*, 2001, **97**(24), 12970–12975.

29. T. Eisner and D. J. Aneshansley, Spray aiming in the bombardier beetle: photographic evidence, *Proc. Natl. Acad. Sci.*, 1999, **96**(17), 9705–9709.

30. D. Kwon, One woman's ability to sniff out Parkinson's offers hope to sufferers, *Sci. Am.*, 2015, available from https://www.scientificamerican.com/article/one-woman-s-ability-to-sniff-out-parkinson-s-offers-hope-to-sufferers/. Accessed 12th June 2017.

CHAPTER 3

The Hallmarks of Hulk

ISABEL PIRES

University of Hull, School of Life Sciences, Cottingham Road, Hull
HU6 7RX, UK
E-mail: i.pires@hull.ac.uk

3.1 INTRODUCTION

When Bruce Banner transforms into the Hulk, his body goes
through a series of fantastic changes that one would imagine to
be possible only in a universe where superheroes exist. However,
this chapter discusses how the biological processes behind his
transformation are similar to some of the processes that a normal cell goes through to become that biggest of supervillains, the
cancer cell. Understanding how this process begins, using the
Hulk as our inspiration, we will try to come to terms with why
cancer happens and how it grows, which is the first step to
understand how to target and destroy it.

The Secret Science of Superheroes
Edited by Mark Lorch and Andy Miah
© The Royal Society of Chemistry, 2017
Published by the Royal Society of Chemistry, www.rsc.org

3.2 ALL ABOUT THE HULK

The Hulk first appeared in print as the 'Incredible Hulk' issue #1 in 1962.[1] Here, nuclear physicist Dr Bruce Banner was exposed to gamma radiation from a gamma bomb explosion when trying to save the life of a teenager. The consequence of his exposure was his capacity to rapidly transform into another being, the Hulk, an incredibly powerful, massive creature with super strength. This transformation occurs when Dr Banner becomes angry and the angrier he becomes, the more powerful he gets.[2] This massive bulk and extraordinary strength leads to some other very useful abilities for a superhero. For example, he can move extremely fast, jump astonishing distances and heights, create shock waves when he claps his hands, and his body can resist a vast array of physical injuries. Oh, and he's also green, but we'll come to that later.

So, have you ever wondered how the Hulk's transformation happens, at a biological level? What happens to his cells? How does his body cope with the change? When thinking about these questions, it occurred to me that there were clear parallels between how the Hulk's cells change when he transforms, and how a normal cell becomes a cancer cell.

3.3 WE HAVE TO TALK ABOUT CANCER: A NOTE BY THE AUTHOR

As a scientist and cancer biologist I have always been fascinated by the character of cancer cells. For example, they are different from the normal cells from which they originate in various ways. Also, their biology changes in a way that is different. I was always conscious of how critical it was to have a good understanding of these differences if we are to design newer, better ways to treat cancer and perhaps even cure it. However, I am also fully aware of how the mention of the word 'cancer' can be terrifying, earth-shattering, and difficult to comprehend. So, this section is focused on clarifying some of the key concepts about cancer biology and how they have progressed in the past few decades.

On a most fundamental level, cancer describes an uncontrolled, abnormal growth of cells in a tissue of an organism, which we call a tumour. When this growth is restricted to the tissue where it started, we call that a primary tumour. Furthermore, when cells in that tumour mass start to change in such a way that they stop communicating with their neighbours and become able to move and spread throughout the body and colonise other organs or tissues, then the cancer becomes aggressive and invasive. Most of cancer-associated deaths are linked with these distant tumours, or metastases.

While the presence of cancer is always a deeply troubling diagnosis, more than 50% of cancer patients in the UK survive for more than five years after their initial diagnosis.[3] What's more, we have advanced our understanding of cancer biology enormously, especially in the last few decades. A fundamental moment in the cancer research field occurred in 2000, when a seminal review noted that there are key features that are key characteristics of cancer cells and which allow the tumour to survive and grow.[4] These are called the hallmarks of cancer.

In 1999, two cancer research scientists, Doug Hanahan and Bob Weinberg, went for a walk during break time at a conference they were attending. During this stroll, they mulled on the idea that, during the preceding 15 years or so, the knowledge of what cancer is, what factors control its development, growth and spread, had exponentially increased and continued to do so. From those initial thoughts the 'The Hallmarks of Cancer'[4] was published in 2000, followed in 2011 by its update, 'Hallmarks of Cancer: The Next Generation'.[5] These reviews established ten conceptual characteristics, or hallmarks, that unify what we understand is needed for a fully-fledged cancer to arise.[†] Together, they consist of the following (see Figure 3.1):[4,5]

- genome instability and mutation
- resisting cell death
- deregulating cellular energetics
- sustained proliferative signalling
- evading tumour suppressors
- avoiding immune detection
- enabling replicative immortality

[†] It is important to note that, although these hallmarks do help to understand the overarching characteristics of what makes a cancer, cancer is not a single disease, but a plethora of diseases with these characteristics in common.

- tumour promoting inflammation
- inducing angiogenesis
- activation of invasion and metastasis

These insights are mandatory reading for all budding cancer scientists and they are extremely helpful when trying to make

Figure 3.1   The hallmarks of cancer. These ten key fundamental characteristics of an aggressive tumour. © Andy Brunning 2017.

sense of what happens to the Hulk, when the mutation of Bruce Banner begins.

3.4   IN THE BEGINNING THERE WAS MUTATION!

Think of a superhero or supervillain.[‡] Done? Good. Does their origin story mention a mutation? It is very likely that the answer to that question is yes. Mutations, as the starting point for super-powers, is a trope of superhero stories, even if you exclude heroes and villains who are born mutants, such as within the X-Men universe.

The same is true for the majority of cancers which, with the exception of hereditary (or family-linked) cancers, arise from mutations. However, in both cases, before mutation there had to be damage to DNA – the molecule that carries all of our genetic instructions.

Now, not all damage to our DNA is significant enough to cause a mutation. In fact, our bodies receive thousands of lesions to the DNA every day, such as small nicks, bumps or cuts. Indeed, we receive damage to our DNA just by being alive. Products of cellular respiration (the process by which we obtain energy from sugars, which uses oxygen), UV light when you go for a walk in the sun, mistakes in DNA duplication during cell division, amongst others, all can cause DNA damage, as do other external or environmental sources, such as by-products of smoking, numerous chemicals, and radiation of several types. Luckily for us, our cells have extremely efficient molecular machines that repair this damage.[§] However, when these lesions cannot be repaired, temporary changes can become permanent and can therefore become a mutation. When this change in the genetic material of a cell occurs, and is capable of being passed on to its daughter cells as a permanent feature in the genome, we call this a mutation.

In cancer progression, mutations to key genes that encode for proteins that repair DNA or regulate cell suicide (a process called apoptosis) make that cell more likely to gain even more

---

[‡]There are more than 10 000 of these, so take your time.
[§]The 2015 Nobel Prize for Chemistry was awarded to Tomas Lindhal, Paul Modrich and Aziz Sancar for their 'mechanistic studies of DNA repair', showing how important these processes are.

mutations. This characteristic is called *genomic instability* and it is a key hallmark and driver of cancer. The loss of apoptosis, or 'resisting cell death', is also another key hallmark, and is particularly difficult to counteract when trying to make effective cancer-targeting therapies.

So, how does this all link back to the Hulk? In his origin story, Bruce Banner was exposed to really high doses of gamma radiation.[1] However, instead of killing him, it led to him being able to transform into the Hulk. Gamma radiation is a high-energy type of radiation that is able to ionise atoms, which, in biological terms, is really bad news. Ionising radiation is able to cause several different types of DNA damage, including breaks to both strands of the double helix simultaneously, which we call double strand break. This is the most lethal type of DNA damage, and just one double strand break can lead to a cell's death. What is really important about these double strand breaks is that, on top of simpler mutations to the genetic code, they can lead to complete rearrangements of whole chromosomes, such as loss of parts of chromosome arms and exchange of material between chromosomes. For cancer cells, all these mutations and chromosomal rearrangements underpin the changes underlying the hallmarks of cancer, as protective genes are repressed, and pro-growth and pro-metastatic genes are made more active.

In the case of Bruce Banner, the dose of gamma radiation he would have been exposed to would cause extensive and catastrophic damage to the DNA, DNA damage and chromosomal aberrations. There is actually a process called chromothripsis – which means chromosomes shattering.[¶] Chromothripsis is an extreme form of chromosomal rearrangement, involving simultaneous breaks on one or more chromosomes.[6] It is thought to be caused by a catastrophic event, such as exposure to gamma radiation. Chromothripsis was initially proposed as a possible origin for the Hulk's initial transformation after radiation exposure by Stanford University scientist Dr Sebastian Alvarado[‖].[7]

[¶]The term derives from the Greek 'chromos' (pertaining to chromosomes, which are in turn named after the colour they take up after being stained with specific chemicals in the lab) and 'thripsis' (shattering into pieces).

[‖]Dr Alvarado will feature a couple more times in this chapter, as the best-known 'Hulk biologist'.

But importantly, as Bruce Banner survived the ordeal, these temporary alterations to his DNA sequence and organization led to permanent changes in his biology that gave his cells at least some cancer-like proprieties, *via* the power of mutation and genomic instability. The first of these is the ability of the changed cells to divide fast, well beyond normal limits.

3.5 GROWTH WITHOUT END

Two of the hallmarks of cancer are directly linked with the regulation of how much and how fast cells divide. These are 'sustained proliferative signalling' and 'evading growth suppressors'. Normal cell division, also known as the cell cycle, is a tightly regulated process that can only occur if the conditions surrounding and within the cell are just right. Understanding this process is so important that the scientists who discovered the elegant molecular switches that regulate cell division were awarded the Nobel Prize for Physiology or Medicine in 2001. The process of cell division is safeguarded by proteins that ensure that a cell only begins to divide if it is in a fit state to do so, and that all of its genetic material is intact. One such protein is a dainty little thing called p53, also known as the guardian of the genome.[8] It is the quintessential tumour suppressor protein, because its main role is to prevent normal cells becoming cancer cells. It does this in several ways, including stopping the cell cycle progression if DNA damage is present. It can even make a cell that is not fit for purpose, or which is at risk of becoming cancerous commit cell suicide.** Cancer cells are very likely to have acquired mutations in tumour suppressors such as p53 (in fact, more than 50% of all solid tumours do exactly that). This allows them to evade these normal control mechanisms and to pass on cancer-promoting mutations to its daughter cells after cell division. This will promote even more out-of-control growth and an escalation of the tumour development process, which is very bad news indeed.

Another way cancer cells can grow out of control is by increasing their sensitivity to growth-promoting signals. This can occur

---

**There are many tumour suppressors in various other cellular processes. DNA repair proteins are also tumour suppressors. However, to discuss them all would require a whole new book...

in the guise of an increased number of receptors for molecules, such as hormones and growth factors. Alternatively, it can occur by hyper-activation of proteins that regulate pro-growth signals, such as those driving the cell cycle forward. Again, this is mostly driven by mutations. In this case, the mutations happen to genes that are normally 'off' or are in a low-activity form, which can then become abnormally 'on' or become hyper-activated. We call these 'oncogenes' because they promote cancer cell growth and survival, and they are the villains to the tumour suppressor heroes.

All of this is incredibly useful to know when trying to make sense of what happens when Banner turns into the Hulk. It tells us that during the bombardment with gamma radiation it is possible that several oncogenes and tumour suppressors in Bruce Banner's cells become mutated, which led to his capacity to change. In other words, his cells – such as cancer cells in a similar situation – no longer respond to 'off' switches and/or have permanently 'on' signals, and can, therefore, divide faster, allowing for that incredible increase in size and bulk during his transformation.

It is sometimes thought that, since his exposure to the radiation, Bruce Banner – and the Hulk – appear to have become ageless and able to regenerate his tissues easily. A key hallmark of cancer that might be underpinning this is 'replicative immortality'. One of the facts that stun most people about tumours is that, given enough space and food, cancer cells are virtually immortal. This is due to the increased presence and activity of a very specialised protein, telomerase.

Whenever a cell divides, its telomeres become shorter and, as a result, their new length determines how many times it can divide. This maximum number of cell divisions is called the Hayflick limit and, on average, it is between 40 to 60 cell divisions for normal cells when in culture in the lab. However, cancer cells are able to keep their telomeres long. They do this *via* telomerase and we know this thanks to Elizabeth Blackburn, Carol Greider and Jack Szostak, who won the Nobel Prize for Physiology or Medicine in 2009 for this discovery. Telomerase adds repetitive DNA sections (telomeres) at the end of chromosomes. Abnormally high levels and/or activity of telomerase are a key feature of cancers. The upshot of this is that they have the capacity of replicative immortality, or the ability to just keep on dividing. This is incredibly useful when you are a superhero who grows to a massive size and who needs to regenerate masses of cells in one go.

However, one problem with rapidly dividing cells and tissues is that they become very fuel and oxygen hungry. So, what needs to happen next to sustain the Hulk's (and a tumour's) raging growth?

3.6   HUNGRY GASPING CELLS!

Let's simplify our model for a moment. In the lab, it is possible to grow small balls of cancer cells – called spheroids – that mimic how a very simple tumour grows and changes. We do this by isolating cells from a bit of a patient's tumour tissue, and grow them until they carry on dividing indefinitely. We call these cancer cell lines. The first one to be established, as well as the most famous and widely used one, is called HeLa. These cells were named after Henrietta Lacks, the patient from which they were originally isolated back in the 1950s. Yes, you read that correctly, over 60 years ago. Now, that is a real superpower![††]

Now, imagine that this very small ball of cells is placed in a vessel containing media that provides nutrients and oxygen. In this initial scenario, all cells in the sphere will receive similar amounts of nutrients and oxygen because the nutrients and oxygen can penetrate inside the small ball of cells. However, as cells get their food and oxygen, they continue to grow. Also remember, these are cancer cells which divide fast. So, within a day or two (depending on the cell type we chose to start off with), the ball of cells is now twice as large as the original one. If we say that the original sphere had a diameter of about 0.2 millimetres, it is now nearly half a millimetre wide. This is a problem, because the oxygen diffusion limit is around 0.15 millimetres, and the cells in the middle the ball of cells have used up all the nutrients and oxygen they had originally available. In these laboratory

---

[††]There are few more interesting facts about the fascinating HeLa cells.[9] All the HeLa cells ever grown in labs since they were first isolated roughly correspond to 50 million metric tons, or about the size of 100 Empire State buildings. For more information I highly recommend the fascinating book *The Immortal Life Of Henrietta Lacks* by Rebecca Skloot.

conditions, our ball of cells of about half a millimetre will therefore contain a low-oxygen, or hypoxic, core.[10]

As you would expect, when it comes to cancer, things are not as clear cut. Tumours are not perfect spheres. Therefore, regions of low oxygen are present throughout the bulk of the majority of solid tumours in an irregular fashion, and, if no oxygen becomes available, they will die within a few days. A rapidly growing tumour mass expands away from the pre-existing blood vessels, and therefore nutrients and oxygen cannot reach the inside cells.

One of the original hallmarks of cancer from the 2000 review was the induction or activation of a process called angiogenesis. This is a term, derived from the Greek words 'angio', meaning vessel, and 'genesis', meaning origin. However, this is a misleading term. Angiogenesis is actually the name for a process by which new blood vessels sprout from pre-existing ones, so it needs a pre-existing vasculature. This is a normal physiological process that occurs in embryonic development, wound healing, and as part of a woman's reproductive cycle. Cancer cells, as you would predict, exploit this aspect of normal biology to their advantage. They activate angiogenesis through a switch-like process that is turned on by low oxygen conditions. This process allows blood vessels in the tissue next to the tumour to grow into it, to provide it with nutrients and oxygen. This means that the levels of oxygen inside a tumour can fluctuate really fast, as areas of rapidly growing cells deplete oxygen and lead to hypoxia, which in turn switches angiogenesis on and reoxygenates that area.

To make matters worse, a lot of reactive oxygen species can be formed during this reoxygenation process, which means more damage to the DNA and an acceleration of tumour development. As well as triggering a complete shift in tumour biology and promoting/activating all the hallmarks of cancer, the absence of oxygen can also decrease the effectiveness of radiotherapy and sometimes chemotherapy, as many of these require the presence of oxygen.

We can therefore predict that the Hulk's muscles have extensive angiogenesis happening all the time, to sustain all the oxygen and nutrient demands of his mighty Hulk cells. However, there is another explanation for how the Hulk's biology might have adapted to increased oxygen demands inherent to his huge bulk, and this is actually found in-universe. The explanation is described in Hulk #77 (2005) where, after fighting a giant squid, the Hulk reflected on how he just survived such a prolonged battle underwater.[2] He comes to the conclusion that his body has adapted to allow him to breathe underwater *via* a new specialized gland that produces an oxygenated perfluorocarbon emulsion, allowing him to breathe and fight underwater. Perfluorocarbons (PFCs) are highly stable molecules in which relatively high volumes of gases, such as oxygen, can be dissolved.

Although the function of the Hulk's PFC gland was originally proposed as arising from his need to survive an underwater battle, the concept could also apply to a large muscle mass with potential regions of hypoxia. Increasing oxygenation levels of tumours have been proposed for many years as a way to improve tumour response to therapy. In fact, 'artificial blood' or 'synthetic blood' perfluorocarbons were proposed several years ago as oxygen carriers to increase oxygen delivery to tumour tissue. However, although successful in laboratory settings, these have not yet been tested in clinical trials.[11]

Low levels of oxygen and/or high-energy demands also trigger other changes in a cancer cell's biology. In normal conditions, cells obtain energy from sugars through a process called oxidative phosphorylation, or OXPHOS for short. This process uses oxygen as a catalyst, and that why it is called aerobic (or oxygen-dependent) respiration. When oxygen availability is low, the cell still needs to obtain energy, and one way it can do this is to switch to anaerobic, or oxygen-independent, respiration.

As with many of the processes hijacked by cancer cells, this is another normal physiological process. In fact you will have experienced it yourself when feeling 'muscle burn' during exercise. This sensation is a consequence of anaerobic respiration. In the absence of oxygen, the normal processes of OXPHOS switch to a simpler, less energetically efficient process called glycolysis. When this happens, an excess of glucose is taken into the cell, and processed to produce molecules that can be used as energy currency. A consequence of this is the accumulation of lactic acid, or lactate, in the spaces between the cells. This is what causes that burning sensation, and why tumours can become very acidic. So, it's no wonder that the Hulk gets angrier when he transforms: all that quick build-up of lactic acid must really make his enormous muscles sting! This altered metabolism is another hallmark of cancer and it is also promoted by low-oxygen biology.

3.7 CHANGING BACK TO BRUCE BANNER: THE POWER OF GENETIC SWITCHES

We have established how Bruce Banner's cells have been permanently changed by radiation, and how this can lead to his transformation to the Hulk. But, how does he change back? Dr Alvarado has also postulated that the reversibility of the Hulk's transformation is associated with an epigenetic switch triggered by external stresses. Epigenetics is a field of genetics that focuses of the regulation of the activity of genes beyond their sequence. In epigenetic processes, molecular tags are added to the *promoters* of genes, which are sections of DNA at the start of a gene. Proteins that induce the process of protein expression from genes bind to gene promoters. This can turn genes 'on' or 'off', depending on specific cell and environmental conditions. Importantly, these are reversible, unlike mutations or chromosomal rearrangements. Moreover, cellular stress can promote these epigenetic changes. So, it is plausible that Banner's anger triggers the Hulk's transformation. In fact, a 2015 study showed that an epigenetic change to the receptor of oxytocin led to alterations in the perception of fear and anger.[12] This is important, because oxytocin is a molecule known to be key for the regulation of social and emotional behaviours. Epigenetic switches have also been extensively shown to be linked with the promotion of several of the hallmarks of cancer discussed in this chapter.

There is also extensive evidence linking stress-associated hormones such as cortisol and adrenaline to several of the hallmarks of cancer, including immune evasion, angiogenesis, invasion and increased inflammation.[13] Two of these, 'tumour-promoting inflammation' and 'immune system evasion', are some of the more recently added hallmarks. Both are further examples of normal cellular processes hijacked by tumours to promote their growth and spread. Inflammatory responses are normally deployed in response to infection or during wound healing. However, it has been long recognized that the tumour environment is characterized by the occurrence of inflammation.

It is possible that similar processes occur during the fast, biologically challenging, transformation of the Hulk, as protective mechanisms. However, there is no evidence that directly links increased cell division with stress and anger hormones, so the biology behind how anger directly triggers the increase in size and bulk during the transformation of Bruce Banner into the Hulk remains unclear.

3.8   IT IS EASY BEING GREEN!

One thing we have not discussed yet is the Hulk's colour, and how this is linked with this whole narrative regarding cancer biology. When Bruce Banner transforms into the Hulk, his body, as well as becoming enormous and super strong, also becomes green.[‡‡2] One theory on how this happens was put forward by Hulk science guru Sebastian Alvarado in 2014. He discussed that the Hulk's green-ness is due to the extensive damage to the muscle tissue, and subsequent accumulation of by-products of bruising.[7] The biology behind this is actually quite interesting and is based on the fact that, during the transformation, a lot of red blood cells are destroyed, and this leads to an accumulation of pigments, such as biliverdin, which is a by-product of the breakdown of the heme group of haemoglobin (Figure 3.2). Biliverdin gives bruises their characteristic greenish hue. The process of breaking down haemoglobin by the enzyme HO-1 (heme oxygenase-1) and the products of this reaction, such as biliverdin and CO (carbon monoxide), have been associated with the promotion of some of the cancer-promoting characteristics we have discussed so far.[14] Specifically, they have been linked with the induction of increased cell proliferation (leading to increased growth) and resistance to cell death. In fact, HO-1 has been identified as a potentially interesting new anti-cancer therapy target.[15]

However, there is another side to this process of going green: biliverdin and another by-product of heme destruction, bilirubin, also have anti-mutagenic proprieties through their anti-oxidant effect, and could be a Hulk-specific strategy to protect his cells.[16]

During the fast, dramatic increase in size and bulk during the transformation, a body-wide inflammatory response is very likely to occur. It has been described that, for cancer, the other rapidly growing tissue we have been talking about,

---

[‡‡]In his original appearance in The Incredible Hulk #1, the Hulk was actually grey.

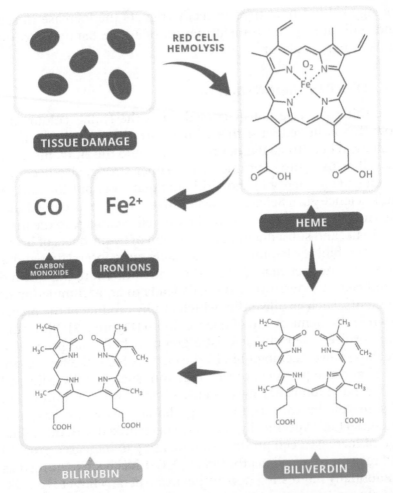

Figure 3.2 Breakdown of haemoglobin. When tissue damage happens, red blood cells disintegrate, a process called hemolysis. During this process, the haemoglobin is broken down into globin (its protein component), iron ions and heme, with the release of carbon monoxide (CO). Heme breaks down into biliverdin, a green pigment. Biliverdin then gets converted to bilirubin, an orange-yellow pigment. It is these pigments that give bruises their characteristic green-yellow tone. © Andy Brunning 2017.

inflammatory cells produce reactive oxygen species, which can cause damage to DNA. Therefore, the production of biliverdin (and bilirubin), one consequence of the massive fast expansion of the Hulk's mass, can protect his cells from damage to DNA from this extensive inflammatory response. This way, biliverdin could protect the Hulk's DNA from permanent damage and, in the end, prevent the escalation of further genetic changes or mutations.

3.9   SOME FINAL THOUGHTS ...

From this discussion, the biology of the Hulk closely resembles the biology of a cancer cell. However, he is not a walking, talking, jumping, smashing, tumour. One key difference we already discussed is that the Hulk's transformation is reversible and, to a degree, controllable. Also, as we discussed in the previous section, some of the biology of the Hulk protects against some of the cancer-like properties of his cells. But, importantly, there is one key hallmark we have not yet discussed that is arguably the most lethal of them all, at least from a patient's point of view: 'activation of invasion and metastasis'. This hallmark covers the ability of cancer cells to alter in such a way that they can escape the primary tumour, invade adjacent tissue and the vasculature, survive the journey through the blood and lymph, and colonize and grow in another organ in the body. This overarching process, also called the metastatic cascade, ultimately causes the majority of cancer deaths.

So, why aren't the Hulk's cells, which are subjected to cancer-forming and cancer-promoting stresses such as genomic instability, inflammation, angiogenic and metabolic pressures, replicative immortality and rapid growth, affected by this? Perhaps it is because this happens in a controlled environment, or perhaps because it occurs in a universe where men and women can fly, can run at the speed of light, and can control the weather.

And, just in case you are wondering after all this Hulk scholarship, I still don't understand how his trousers stay nearly intact when he transforms.

REFERENCES

1. S. Lee and J. Kirby, *The Incredible Hulk #1*, Marvel Comics, 1962.
2. Marvel, *Marvel Universe Wiki: Hulk (Bruce Banner)*, 2016 [last accessed 23/10/2016], available from: http://marvel.com/universe/Hulk_(Bruce_Banner).

3. UK CR, *Cancer Survival Statistics*, 2016 [last accessed 23/10/2016], available from: http://www.cancerresearchuk.org/health-professional/cancer-statistics/survival.

4. D. Hanahan and R. A. Weinberg, The hallmarks of cancer, *Cell*, 2000, **100**, 57–70.

5. D. Hanahan and R. A. Weinberg, Hallmarks of cancer: The next generation, *Cell*, 2011, **144**, 646–674.

6. P. J. Stephens, C. D. Greenman, B. Fu, F. Yang, G. R. Bignell and L. J. Mudie, *et al.*, Massive genomic rearrangement acquired in a single catastrophic event during cancer development, *Cell*, 2011, **144**, 27–40.

7. S. Alvarado, *Stanford Researcher Explains the Science Behind the Incredible Hulk*, 2014 [last accessed 23/10/2016], available from: https://http://www.youtube.com/watch?v=cwzpgHyq65s.

8. S. Armstrong, *P53: The Gene that Cracked the Cancer Code*, Bloomsbury Publishing, London, 2014.

9. R. Skloot, *The Immortal Life Of Henrietta Lacks*, Crown Publishers, New York, 2010.

10. J. Friedrich, R. Ebner and L. A. Kunz-Schughart, Experimental anti-tumor therapy in 3-D: Spheroids – old hat or new challenge? *Int. J. Radiat. Biol.*, 2007, **83**(11–12), 849–871.

11. C. I. Castro and J. C. Briceno, Perfluorocarbon-based oxygen carriers: review of products and trials, *Artif. Organs*, 2010, **34**(8), 622–634.

12. M. H. Puglia, T. S. Lillard, J. P. Morris and J. J. Connelly, Epigenetic modification of the oxytocin receptor gene influences the perception of anger and fear in the human brain, *Proc. Natl. Acad. Sci. U. S. A.*, 2015, **112**(11), 3308–3313.

13. Y. Gidron and A. Ronson, Psychosocial factors, biological mediators, and cancer prognosis: a new look at an old story, *Curr. Opin. Oncol.*, 2008, **20**(4), 386–392.

14. M. D. Hjortsø and M. H. Andersen, The expression, function and targeting of haem oxygenase-1 in cancer, *Curr. Cancer Drug Targets*, 2014, **14**(4), 337–347.

15. L. Y. Chau, Heme oxygenase-1: emerging target of cancer therapy, *J. Biomed. Sci.*, 2015, **22**, 22.

16. A. C. Bulmer, K. Ried, J. T. Blanchfield and K. H. Wagner, The anti-mutagenic properties of bile pigments, *Mutat. Res., Rev. Mutat. Res.*, 2008, **658**(1–2), 28–41.

CHAPTER 4

Supervillainy 101: Choosing Between a Zombie, Vampire or Werewolf Apocalypse

J. VERRAN*[a] AND M. CROSSLEY[b]

[a]School of Healthcare Science, Faculty of Science and Engineering, Manchester Metropolitan University, Chester Street, Manchester M1 5GD, UK; [b]School of Computing, Mathematics and Digital Technology, Faculty of Science and Engineering, Manchester Metropolitan University, Chester Street, Manchester M1 5GD, UK
*E-mail: j.verran@mmu.ac.uk

4.1 AN INTRODUCTION TO EPIDEMIOLOGY

More often than not, superheroes are faced with the threat of the world ending. These apocalyptic scenarios, whether natural, brought about by the antagonist, or humanity's own shortcomings, are often the primary threat of the story and, although the nature of the threat varies, there is no more terrifying scenario than that of a disease wiping out the whole of humanity.

So, let's imagine we are a budding archvillain. We have decided that we are going to bring about the end of humanity – as

The Secret Science of Superheroes
Edited by Mark Lorch and Andy Miah
© The Royal Society of Chemistry, 2017
Published by the Royal Society of Chemistry, www.rsc.org

supervillains often do – with an infectious agent. ACME Diseases Inc. have a sale on, and are offering three infectious diseases at a very reasonable price:[†]

- 'Zombification' – a fairly standard zombie transmission and resultant apocalypse
- 'Vampirism' – a well-known, but previously thought to be fictitious, disease
- 'Lycanthropia' – the disease that causes an individual to live out the remainder of their days as a werewolf, cursed to transform into a vicious, raging beast whenever the moon is full

Now we need to decide which choice is most likely to achieve what we want, which is to wipe out or enslave all of humanity (except ourselves of course!), and then we enter our credit card details, and start planning ...

We're going to take a little tour of epidemiology, considering each of our potential diseases and comparing them to real-world diseases. As we go, we're going to score each of our possible options out of five, then tot up the scores at the end to see which of the diseases we should actually order.

To start, we need a bit of background, so that we can make an informed choice. Worldwide, infectious diseases already account for 23.5% of 58.7 million annual deaths.[1] Although in the twentieth century there was a significant downturn in

---

[†]Have you ever thought it odd how supervillains seem to have largely ignored vampires, werewolves, zombies and their ilk when considering ways to wipe out civilization? After all you'd have thought the villainous might have taken some inspiration from the classic gothic tales and be quite keen on the undead. This lack of walking dead in comic books is actually the result of the Comics Code Authority (http://cbldf.org/the-comics-code-of-1954/), a voluntary regulatory body founded in 1954, that set out to make sure this new cultural medium didn't corrupt the youth of the age (does that sound at all familiar?). The code included such gems as 'Policemen, judges, Government officials and respected institutions shall never be presented in such a way as to create disrespect for established authority', 'Passion or romantic interest shall never be treated in such a way as to stimulate the lower and baser emotions.', 'Inclusion of stories dealing with evil shall be used or shall be published only where the intent is to illustrate a moral issue ...' (*i.e.* the goody always wins) and of course 'Scenes dealing with, or instruments associated with walking dead, torture, vampires and vampirism, ghouls, cannibalism, and werewolfism are prohibited.' It wasn't until 2001 that Marvel abandoned the code and DC didn't do so until 2011!

morbidity and mortality associated with infectious disease, we witnessed the emergence of new infectious diseases such as HIV/AIDS, Legionnaires disease, 'mad cow' disease – and many others. In addition, re-emergence of diseases previously deemed to be 'under control', such as tuberculosis, was observed. Globally, HIV/AIDS, malaria and tuberculosis remain problematic; the influenza H1N1 pandemic (2009) caused worldwide concern; and the World Health Organization (WHO) has also designated neglected diseases as a focus, due to their impact on the overall health of populations in developing countries.[2] In 2016, we can add antimicrobial resistance, Ebola and Zika to the catalogue of fears around infectious disease. So, real infectious diseases have been pretty successful in the past. What can our monsters do that is different, or more effective?

Epidemiology is the study of the disease in populations. Epidemiology deals with public health rather than the health of the individual. The microbial pathogen that causes the infection is transmitted to an uninfected, non-immune individual *via* living carriers (when the carrier is obviously infected – or even not obviously infected at all) or *via* inanimate objects, often called 'fomites' – things like chopping boards, towels, toothbrushes, pens, computer keyboards, telephones (dirty phones are a particular worry, they did after all wipe out the Golgafrinchans[3]), *etc.* The microbial pathogens cannot usually multiply on these inanimate surfaces without food and water, although they can survive for periods of time. Vampires and zombies are often referred to as the 'undead', or the 'walking dead' respectively. We might argue whether or not they are 'animate', but they certainly can pass on the characteristic 'symptoms' which they display (which a contaminated telephone certainly cannot!).

Looking at some of the terms used in epidemiology, 'epidemic' is probably the most familiar. This describes an unusually high number of infected individuals in a population, typically linked by time and space, perhaps from a common source (for example a banquet, or the water supply), or by host to host spread. 'Pandemics' are widespread, usually worldwide epidemics. The term 'outbreak' describes an incident where two or more cases are linked by time and space, often in an area where only sporadic (occasional) cases are observed. 'Endemic' diseases are constantly present, usually at low incidence, in a population.

Individuals infected with an endemic disease are reservoirs of infection.

So, zombies tend to cause pandemics, vampires are associated with outbreaks, and werewolves are typically present constantly but sporadically. This might already make a zombie outbreak seem the tempting choice for a budding supervillain – why choose a disease with potentially only a local scale, when you can go global? It's not quite that easy, though – and we can investigate further by considering the properties of each of our three monstrous diseases.

Examples of 'real world' pandemics are influenza and plague. The important key question is 'What features enable pandemic spread?' For influenza, it is high infectivity and a short incubation period. (For example in the H1N1 pandemic of 2009, the infection rate was high, but the mortality rate was only 0.02%). For plague, it was poor living conditions, with people living in close proximity to one another and to rats – and their fleas. The fleas passed plague onto humans by puncture (bites); subsequently pneumonic plague helped human-to-human transmission through inhalation. The mortality rate of pneumonic plague approached 100% prior to the use of antibiotics, but the transmission rate can be controlled by careful handling of patients.[4] In the case of zombies, 100% transmission and 100% infection rates are key features of spread, aided of course, by the unstoppable nature of the infected agents (the zombies!).

Another consideration is the incubation period. The average incubation period of a zombie infection varies significantly, depending on the source of information. In the movie *World War Z*, it is a mere 12 seconds (which does not mirror anything seen in the real world of disease), or – for 5% of the population – a little longer (10–15 minutes, also unfeasibly short).[5] However, in the movie, the short incubation period and the speed with which the zombies move lets us watch the spread of infection almost in real time, although we are required to suspend belief frequently as the plot progresses. In the novel *World War Z*, on which the movie was based, the incubation period extends to days, which is more realistic, and gives time for the pandemic to emerge on a global scale.[6] Many older zombie information sources, such as the classic Romero movie adaptations,[7–9] where the zombies are the more traditional slow, shambling, type, also present this slower incubation period. This longer period enables suspense in the narrative, where the infected are camouflaged before turning. With SARS (sudden acute respiratory syndrome), with transmission by inhalation, the incubation period during the 2002 outbreak (typically 2–7 days, even up to 10 days)[10] was long enough that air travel enabled disease distribution around the world before symptoms of infected individuals were recognized. Person-to-person transmission is essential for the spread of the zombie condition, except for rare occasions of transplants and transfusions,[6] revealing progressive and coherent spread. The 2014 Ebola outbreak similarly progressed slowly but inexorably, relying on person-to-person contact, with occasional cases leaving infected areas *via* air travel (*e.g.* to the US, Spain and the UK). There were individual cases of transmission in those countries, but outbreaks did not occur because of significant cross-infection control. In that outbreak, the fatality rate was around 50%, whereas in the 1976 Zaire outbreak it approached 90%.

The ability of vampires to move rapidly, and at night, enables the emergence of geographically isolated clusters of cases, best fitting the outbreak scenario. Often there is a 'parent' vampire, whose victims tend to remain associated with (usually) him. In recent times, vampires have tended to be extremely attractive and seductive, which helps them to lure their targeted prey. Sexually transmitted diseases similarly can present as outbreaks, and epidemiologists try to map contacts in order to appropriately treat them, and prevent spread. This is particularly important with the emergence of antibiotic-resistant strains of bacteria.

Werewolves are few and far between, but are occasionally found in packs.[11,12] Being infectious only once in a month (at the full moon), and being human for the rest of the lunar cycle, means that werewolf condition could be deemed to be endemic, with sporadic cases, and – rarely – outbreaks. This is a typical presentation of an endemic disease. Examples would be cholera in India, or measles in the UK.

We can get a glimpse into the epidemiology of our monsters by looking at the same sort of techniques epidemiologists use to model real infectious diseases. SimZombie[13,14] is an agent-based modelling programme, designed specifically to model the outbreaks of zombies, vampires and werewolves. In results we note typical outbreaks, described by SimZombie, for each of the three monsters (Figure 4.1). The zombies, as described before, are inexorable, spreading out across the entire population from the location

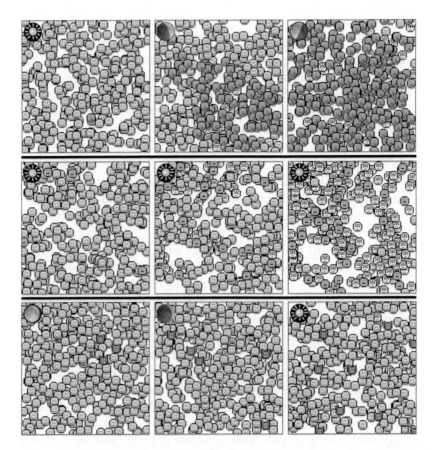

Figure 4.1 SimZombie demonstrates the typical spread of each of the three monsters. Zombies (top row) are an inexorable pandemic. Vampires (middle row) are more selective, and have a tendency to kill their prey rather than turn them. Werewolves (bottom row) are rarely active, and as such, any patterns are much slower to emerge.

of the initial outbreak. The vampire infection has a tendency to leave many more people dead than infected, but is similarly disastrous. The werewolf infection is much, much slower – both in terms of number of casualties and the number of new cases.

So, how are our monsters doing here? As they're the most likely to cause a global outbreak (a pandemic), zombies have to be given a whopping *five out of five* on the apocalypse scale. Affecting far fewer people, and tending to be closer defined as epidemics, or even outbreaks, vampires score *three out of five*, and finally werewolves, our endemic infection, score a measly *one out of five*.

As a budding supervillain, you are probably already thinking that zombies sound like a winning choice, because they're the monster most likely to cause a pandemic – and we know that transmission and infection rates are key to this – but to learn more, we need to think about what actually causes infections.

4.2   THE LIFECYCLE OF DISEASE

Diseases are caused by pathogens – specific types of microorganism[‡] that causes disease. Usually these are bacteria or viruses. You could argue, if you were the arguing sort, that it is these microorganisms that are the true supervillains. We don't want to allow ourselves to play second fiddle, but we have to pay tribute to these tiny monsters that will be doing most of our villainous work.

To cause disease, a pathogen must grow and reproduce in the host. In many cases, an individual pathogen cannot grow outside the host: if the host dies, the pathogen also dies. Pathogens that kill the host before they move to a new host will become extinct. Most host-dependent pathogens therefore adapt to coexist with the host.[1] Although some literature attributes the zombie condition (in particular) to the result of some infection (*e.g.* Rage,[15] Solanum,[6] *etc.*), it is essentially that the monster itself is the visible (non-living) embodiment of infection – and with an undead mobile host, the infection can continue to spread.

Disease progression in an individual follows a series of phases: infection, incubation, the acute phase, decline and convalescence. Infection occurs when the pathogen enters and colonizes the host; incubation is the period between infection and appearance of symptoms. This is short in influenza and long with HIV, and varies with zombies, vampires and werewolves depending on the literary source – as we have already discussed. During the acute period, the disease is at its height. For zombies, this phase does not decline.[§] For vampires, this phase is recurrent (every night), but vampires tend to be able to control their behaviour, being sentient in contrast to the insentient zombie. For werewolves, the acute phase is a violent and frenetic period that cannot be controlled by the sufferer; they anticipate the relapse and often try to restrain/contain themselves in advance.

[‡]A very, very small living thing.

[§]Assuming, that is, that the zombies are able to sustain their state of 'undeath' indefinitely.

Our supernatural monsters experience somewhat different decline and convalescent phases. Vampires are famously undead and immortal. Zombies, being dead, have been reported to deteriorate and rot, often due to secondary infections caused by saprophytic bacteria† from the soil, such as those causing gangrene. Werewolves, generally speaking, are not immortal in the way vampires are. Remaining 'human' for the majority of their life-cycle, they age as a normal human, and are also susceptible to disease, and accidents.

Some pathogens, such as herpes, remain with the host and cause occasional eruptions. These tend to occur in times of stress (*e.g.* 'fever blisters' 'cold sores', during menses, during brain surgery),[16] and could be likened to our lycanthropic condition. Similarly, tuberculosis presents occasions when cases are 'open' and infectious, and other periods when they are not. For all of our classic monsters, with no decline in symptoms, there will be no convalescence: zombies could eliminate humankind and then eventually rot away (although there is no evidence of this); vampires would likely remain in small groups, and werewolves would need to reproduce as they age, in order to maintain their population. If we were choosing our monstrous disease, perhaps some zombie effectiveness coupled with vampire sensibility might be useful?

'Virulence' is a term used to describe how dangerous a disease might be: this can be illustrated through mortality and morbidity data. 'Mortality' describes the incidence of death and 'morbidity' the incidence of disease. For example, respiratory disease and acute digestive disorders present with high morbidity but low mortality, whereas Ebola has low morbidity and high mortality. A well-designed disease needs to balance these two factors: a disease where mortality is too high might kill the host before anyone else had been infected. A disease where mortality is too low is not perceived to be a threat.

Some diseases are characterized by subclinical infections. These are invisible but immunizing, so are the best type to be infected with, but they are not particularly helpful for those uninfected and unknowing contacts around you. For example, up to 99% of polio infections are subclinical infections – which meant that it spread significantly across communities, particularly in

---

†These are another special type of microorganism that live on dead, or decomposing, matter – yum!

the mid-twentieth century, because infected individuals showed no symptoms and were able to transmit infection. The main route for transmission was faecal-oral, where communal and recreational waters were key to spread. In contrast, for smallpox and measles, for example, there is no subclinical infection: every case presents with recognizable symptoms, making them much easier to avoid. This is also true for our zombies – and for vampires and werewolves when they are in their infectious state. Rarely, sub-clinical infections emerge: in *28 Weeks Later*,[17] individuals carrying the infection for extended periods without presenting symptoms were responsible for outbreaks in areas where the infection had been previously controlled.

So how about the scores? You might suggest that the vampire-like disease would win this round. Immortality is the key here, as it allows the host to perpetuate the infection. This property, coupled with the fact that their symptoms can be well-hidden,[‖] makes vampirism a good choice – so they score the massive *five out of five* on the apocalypse factor. Werewolves, claiming no immortality, nor presenting an extended acute phase, do not have a good chance of propagating in a population, so again, they're down at the bottom with a pathetic *one out of five* – sorry werewolf fans. Zombies are somewhere in the middle. Resolving the debate about whether zombies can continue indefinitely, or whether they rot away, would help us make a definitive answer, but the fact that zombieism is not subclinical means they're going to score a maximum of *three out of five* at the very best.

But there are yet other properties we need to consider. So far, we have looked at disease as an abstract concept, focusing on rates and percentages. We also have to take into account the way in which those diseases are spread.

4.3   DISEASE TRANSMISSION

Disease transmission has a significant influence on the features of its outbreaks. In our villainous minds, we are looking for a method that allows for the greatest, but also the most reliable, distribution of our disease – does that come from a zombie-like, vampire-like or werewolf-like disease?

[‖] Assuming they can continue to come up with good excuses for not attending daytime events.

Transmission of disease typically occurs through four main routes:

- inhalation (someone breathes/coughs/sneezes it out, you breathe it in)
- ingestion (you eat it)
- contact (you touch something contaminated with the pathogen)
- puncture (your skin is pierced by a contaminated item – a needle, splinter, or bite)

All of our monster conditions are normally blood-borne, transmitted by puncture (bite), with live prey – there seems to be an aversion amongst zombies and vampires to eating dead meat (werewolves can eat normally for the remainder of the month). The manic and frenzied symptoms are often likened to those of rabies (*e.g. I am Legend*,[18] *WWZ*[6]): zombieism has also been compared to the long-term degeneration and deterioration demonstrated by diseases such as 'mad cow disease' or Creutzfeldt–Jakob disease, but in those cases, the drive to infect others is absent. However, differences in epidemiology have already been noted. Overall, transmission seems to be 100% effective, with the victims either being killed, infected, or, on occasion with vampires, becoming long term 'food'.[19] Whatever the mode of transmission, the underlying motivation of the monsters is invasive reproduction; self-reproduction being a driver for all living things, including pathogenic microorganisms (the agents of disease), who need to replicate in order to survive.

This sort of direct contact route of infection as employed by our monsters is much less efficient – but easier to control – than that of an airborne disease.** In many stories,[6] the fear that the agent may become 'airborne' is overwhelming. In *The Hot Zone* novel,[20] the Ebola strain under investigation is airborne, but of low virulence to humans, tending to cause subclinical infection. In *The Girl with all the Gifts*,[21] the pathogen is, unusually, a mould, whose aerial spores disperse globally, causing an inevitable apocalyptic

---

**Unfortunately, the airborne diseases ACME Diseases Inc. are offering are outside of our limited budget for bringing about the apocalypse – we want to doom the world, but not at the cost of dooming our bank balance.

pandemic. Airborne infections are much more difficult to control – witness also the fear around the H1N1 emergence and the speed of spread, and the number of emerging novels focusing on different types of influenza.

In terms of when infectivity is at its peak, this varies with different diseases. The zombie is continually infectious and endlessly mobile, the vampire is only active at night, usually restricted in movement during the day (unless immunized to tolerate sunlight),[12] therefore tending to cause localized outbreaks where tracing of contacts is possible. The werewolf, being only active during the full moon, is not infectious at all in between, living a normal life.

With no sub-clinical infections, and with more information becoming available about the epidemiology of a particular disease, it becomes easier to map the progress of an outbreak. This is demonstrated particularly well in the movie *Contagion*,[22] where a pandemic is caused by a virus related to Nipah, transmitted from a bat. For many diseases, there are reservoirs of infection – sites in which infectious agents remain viable, and from which individuals may become infected. Rabies and Ebola are good examples. For vampires, bats could be considered a possible candidate, and for werewolves perhaps wolves themselves. 'Zoonosis' is the term given to diseases that primarily infect animals but are occasionally transmitted to humans. For zombies there seems to be no documented reservoir: the outbreaks derive from a range of different scenarios such as biological warfare, ecological disaster or scientific research. In real life, it is desirable that there is no reservoir for a disease: reservoirs make diseases harder to control.

In life, and in fiction, some microorganisms have been developed to function as biological weapons (for example, anthrax), and a variety of outputs describe the creation of some sort of 'doomsday' pathogen that will reduce global overcrowding, with only the chosen few being saved.[23] We have decided to use pathogens in 'human' form – our zombies, werewolves or vampires – so that they are visible, and we can, hopefully, control what they get up to!

In short, there is no major difference between the three diseases on offer in terms of mode of transmission. Diseases spread through direct contact, rather than airborne transmission, seem to be more controllable. Similarly, diseases with no reservoir and no subclinical infections will help us to control our monstrous outbreak. All middling scores here, with each of our diseases getting *three out of five*. The nature of the transmission of the diseases are all too similar to notice a significant difference – what we really need is an airborne disease, but unfortunately for us, there are none available!

4.4 EMERGING ISSUES AND DISEASE EVOLUTION

As overlord of the ruined world,[††] we should think about what might happen to the diseases we have wrought upon the population in the future. A sensible way to begin is to think about the recent evolution of diseases, and the recent evolution of monsters, and to see if there is any overlap between the two.

The microorganisms causing real diseases undergo evolution – for example, *Streptococcus pyogenes* caused scarlet fever and puerperal fever (childbed fever) in the 1840s and was a major killer of children and post-partum women at the time. Nowadays streptococcal sore throat, caused by the same microorganism, is common but considerably less virulent. Many other diseases also evolve over time, owing to the fact that the microorganisms replicate quickly and are required to adapt to new environments rapidly (for example, one of the main fears is microorganism resistance to known antibiotics).

Vampires too have evolved over recent years, becoming alluring and attractive, working in groups and bonding more with humans – indeed demonstrating more human attributes. Do these features mask their virulence and make them more lethal? Vampires seem to be aware that it is important to sustain the predator–prey relationship to avoid making themselves immortally starving. In the novel *I am Legend*[24] the infected population – who exhibit traits of both zombie and vampire – becomes the norm, and the uninfected protagonist becomes the outsider, and is isolated (and killed). In the 2007 film,[18] the animal-like monsters also demonstrate some evolution: there is increasing tolerance to light, rudimentary relationships being formed, and an intelligent focus on the 'Legend' as target.

Zombies are unattractive, 'brain-dead', shambling creatures driven to attack one victim after another. They do not recognize the fact that exhaustion of their prey will mean their own extinction.

[††]An overlord who is determined to bring down society, yet is still a responsible scientist, of course.

However, according to several sources, they have evolved to move more quickly[5,15] – even being re-named 'zoombies'[25] – which makes the apocalypse inevitable, steering the zombie to an evolutionary dead-end, with no remaining hosts to infect. In contrast, reports have been made referring to more humanistic behaviour. This tends to reduce voraciousness due to attempts to consciously modify behaviour.[26-30] This change helps to make us slightly more empathetic towards the zombie – but we must maintain our guard!

Werewolves are less well studied than the other two conditions, and as such, we have fewer data points to extrapolate from, and fewer ideas about where werewolves have come from, and are going, in an evolutionary sense. Werewolves tend to be isolated individuals. They often try to contain their own infectivity. They retain animalistic properties, and when active, cannot be controlled. There is less opportunity for them to interact naturally with the human population due to their self-imposed isolation. There are emerging descriptions of werewolf packs, but overall the small number of cases, the rare opportunities (full moon) for activity, and their own self-control tends to lead them to an evolutionary dead-end.

In general, it appears that zombies have become more threatening to the general population in their recent incarnations and, as such, may continue to develop into more fearsome creatures. This is perhaps not what we are looking for if we want to remain in control[‡‡] – too risky for us to rely on – earning them *two out of five* on the apocalypse scale. Vampires, although more attractive and interactive with humans, maintain their general fearsome personae and essentially ruthless characteristics, and have even demonstrated some resilience to their previous weaknesses.[11,31] This seems particularly promising, and could even mean the vampires can be controlled and guided towards our aims and purposes – *four out of five* for vampires, here. And if we want something reliable, we know that werewolves are stagnant in the evolutionary pool, so we at least know that the disease we are buying is the disease we will be getting (and would continue to get!). A medium score for the werewolves, *three out of five*. While they do not have much going for them here, they are at least reliable!

---

‡‡Worse still, if they follow the other recent trend of becoming more human-like, and even co-existing with the human population, we might end up outnumbered by both humans and our new zombie-children. Definitely not desirable.

4.5   PREVENTION AND CONTROL

Pesky superheroes or, more likely, pesky health professionals, are going to try to control or even prevent our infectious disease from wiping out humanity. An important step in the road to identifying the perfect world-ending infectious disease is to consider how each of the diseases might be stopped. After all, if it's easy to stop our disease, surely we'd be better off choosing a different one?

Epidemiological study of different disease outbreaks enables identification of its source ('patient zero') and transmission. Surveillance helps epidemiologists to identify any changes in incidence of particular diseases, or to note emerging and re-emerging infections. 'Horizon scanning' is important in terms of rapid detection of any disease emergence. Social media have assisted in more rapid observation of emergence. In turn, this knowledge can help prevention and control strategies to be developed.

In epidemiological terms, decisions are made to protect populations, which might have risks for individuals within that population, but the focus in this case is public health, not individual treatment. In the nineteenth century, the most successful disease prevention campaign was the cleaning of the water supply and the introduction of sewage disposal, alongside improved living conditions.

Meanwhile vaccination and other 'prophylactic' strategies helped to protect populations from endemic disease, and diseases for which particular professions or behaviours might be at risk. Examples are vets and rabies, travellers and whatever is endemic or a risk in the country to be visited, healthcare workers and hepatitis. Likewise, 'a clove of garlic a day keeps the vampire at bay' might work well.

One of the most difficult strategies is to rely on changes in human behaviour (*e.g.* changing sexual behaviour to reduce incidence of sexually transmitted disease, or cleaning teeth regularly compared with fluoridation of the water supply). But introduction of a concerted education campaign could help alter behaviours and slow the spread of the disease, something as simple as the slogan 'Don't invite them in' could reduce the likelihood of succumbing to a vampire infection, but is similarly flawed. Perhaps a villain-implemented outbreak of lycanthropy would be stopped in its tracks simply by nudging people to stay indoors during full moons; something as easy as free streaming of the latest movie blockbusters on those nights might well keep people out of harm's way?

A particular strength and positive attribute of the vampire, zombie and werewolf conditions is that everyone is susceptible to infection. Stories have developed around the idea of isolated communities or groups of individuals who somehow have avoided infection. For example, in Charlie Higson's *The Enemy*,[32] *The Dead*[33] and *The Fear*[34] series people under 14 years of age were immune to the zombie-like infection. In *The Girl with all the Gifts*,[21] again it is children who acquire partial immunity and modified symptoms that enable them to survive.

Isolation has been a common strategy in the avoidance of zombie pandemics: this is separation of the uninfected from the infected – this is not the same as quarantine, where possibly infected individuals are separated from the uninfected. In most cases concerning monsters, these isolation strategies have not been particularly successful long-term.

Screening provides information through cross-sectional studies of populations to identify whether infection is current, or past – and again helps to inform of the epidemiological status of a particular infection. Tests for indicators of infection such as antibodies and other features of the immune response or detection of components of the pathogen itself are developed. For our three monster diseases, the zombie condition is evident to the eye – the duration of the incubation period and the onset of infectivity would need to be known if screening were to have any impact. Vampires similarly tend not to be active during the day, but are known to display particular features such as pale skin, and a neat dress sense. More recently, vampires, as we have identified, are better able to blend in and interact with a population. One possible way of screening for a vampire is to use a mirror – after all, it is said that the undead have no reflection.[§§] Werewolves, in their human phase, have no identifiable symptoms, and no known screening methods. Indeed, many werewolf narratives use the idea that the werewolf cannot be identified amongst a population as the point of suspense in the story.[¶¶]

[§§]Whether this would work in practice or not is another question – also, how would you do it? Just ask them politely to stand in front of a mirror? Maybe you could ask them take a selfie?

[¶¶]And there is also a popular role-playing game based on the concept of identifying which person in a group of players is the 'werewolf' (it's called Werewolf, by the way, if you hadn't guessed).

Culling has been used as a strategy to eliminate a reservoir of infection, but in animal rather than human populations. Examples are culling badgers for bovine tuberculosis; culling cattle for foot and mouth disease (2001); and culling chickens for Hong Kong chicken flu (1997 and 2016). In *World War Z*[6] several culls of zombies are carried out, particularly to destroy any carriers who are not yet presenting symptoms. Actually WWZ tries virtually every prevention method, even an ineffective vaccine (placebo), attributing the pandemic in its early stage to rabies. Brain destruction for zombies, stake in the heart for vampires and silver bullets for werewolves have been reported to have some success in individual cases, but in terms of control of an infected population, this strategy is somewhat limited if the monsters hold the winning card in terms of numbers.

Chemoprophylaxis relies on using chemicals to prevent infection. This strategy tends to be restricted to specific conditions; for example, anti-malaria medications, or specific uses of antibiotics to prevent infections. For vampires, garlic has been proposed as a chemoprophylactic: it is known to have antimicrobial properties and has been used for centuries as a traditional remedy. Religious symbols have been known to help in prevention of infection, with running water also as a barrier to travel but, like garlic, the effect only remains as long as exposure – and it could be risky! Of course, vampires are also killed by sunlight.

It appears, then, that all three of the monster diseases are equally difficult to resolve. There are no known cures for any of the three, and although behaviour change could provide a reasonable solution for some, it is difficult to enact a successful campaign to bring about this change. All three of the monsters here earn a terrifying *four out of five* on the apocalypse scale, with each having some weakness – however small or difficult to exploit they may be.

We now have a lot of information to go on. We've looked at the possible lifecycle of our diseases, how each is transmitted, what might happen to them after they've been released, and now we are ready to total up our scores and see which disease we are going to use to wipe out the human race.

4.6 CONCLUSIONS

We started out with our villainous minds turned to the notion that we needed an infectious disease to help us to control the Earth. Having been offered three choices (between a zombie-like, a vampire-like and a werewolf-like disease), we have meandered our way through the ups and downs of each, trying to decide which to use to take over the world.

And at the end, what do we have? Which is most successful?

The vampire wins the overall scoring – coming out at a grand total of *nineteen out of twenty-five* on the apocalypse scale. The subclinical nature of the vampire, the immortality of the infected, and the disease's evolution to overcome its weaknesses are key factors in its victory here.

Although the zombie is the most successful in terms of numbers, we noted that every case is clinical[35] and so the pandemic can never be hidden. Any superheroes who wanted to stop us in our tracks would have very advance notice that it was coming. This is a large factor in zombies coming second to vampires, scoring a very close *seventeen out of twenty-five* on the apocalypse scale.

Lagging far behind, werewolves only managed to scrape together *twelve out of twenty-five* points. Perhaps there is a reason they are not as popular in literature and media as the other two? In terms of the disease, they are mostly severely hampered by their endemic nature – being 'active' only once a month means that there is little opportunity for the infection to spread.

Perhaps we should also consider our own humanity as part of our decision-making process. The zombie pandemic forces us to face how we would behave in an apocalyptic scenario, confronting very difficult decisions. Vampires are attractive, intelligent, articulate and conscious. They conceal symptoms, enabling controlled spread of the condition without exhausting the non-immune population. Vampire tales allow us to consider prejudice, and difference, because the vampires themselves are often sympathetic characters, with human characteristics – particularly around love. Werewolves have not really captured the public imagination: their animalistic features and tendency to be alone means that they remain, well, alone.

We can learn a lot about ourselves and the spread of infectious disease by looking at monsters that are driven to make more of themselves. In the real world, however, the emergence of new and virulent pathogens spreading across the global village means that many of the new supervillains could be the diseases themselves. Be prepared!

REFERENCES

1. T. M. Madigan, J. M. Martinko, K. S. Bender, D. H. Buckley and D. A. Stahl, *Brock, Biology of Microorganisms*, Pearson Education, Essex, England, 14th edn, 2014.
2. World Health Organization, 2016 [cited 2016 11 14], available from: http://www.who.int/.
3. D. Adams, *The Hitchhiker's Guide to the Galaxy*, Pan Books, UK, 1979.
4. J. L. Koos, Risk of person-to-person transmission of pneumonic plague, *Clin. Infect. Dis.*, 2005, **40**(8), 1166–1172.
5. M. Foster, *World War Z*, Paramount Pictures, United States of America, 2013.
6. M. Brooks, *World War Z*, Duckworth Publishers, Surrey, 2006.
7. G. A. Romero, *Night of the Living Dead*, Walter Reade Organization, United States of America, 1968.
8. G. A. Romero, *Dawn of the Dead*, United Film Distribution Company, United States of America, 1978.
9. G. A. Romero, *Day of the Dead*, United Film Distribution Company, United States of America, 1985.
10. Centers for Disease Control and Prevention (CDC), *Severe Acute Respiratory Syndrome (SARS)*, 2013 [cited 2016 11 13], available from: https://www.cdc.gov/sars/.
11. C. Hardwicke, *Twilight*, Summit Distribution, United States of America, 2008.
12. A. Ball, creator, *True Blood*, [Television Broadcast], Home Box Office, US, 2008.
13. M. Crossley and M. Amos, SimZombie: a case-study in agent-based simulation construction, *Proceedings of the 11th International Conference on Adaptive and Natural Computing Algorithms*, Springer, Lausanne, Switzerland, 2013, pp. 110–119.
14. J. Verran, M. Crossley, K. Carolan, N. Jacobs and M. Amos, Monsters, microbiology and mathematics: the epidemiology of a zombie apocalypse, *J. Biol. Educ.*, 2014, **48**(2), 98–104.

15. D. Boyle, *28 Days Later*, Twentieth Century Fox Film Corporation, United Kingdom, 2002.
16. D. A. Padgett, J. F. Sheridan, J. Dorne, G. G. Berntson, J. Candelora and R. Glaser, Social stress and the reactivation of latent herpes simplex virus type 1, *Proc. Natl. Acad. Sci.*, 1998, **95**(12), 7231–7235.
17. J. C. Fresnadillo, *28 Weeks Later*, Twentieth Century Fox Film Corporation, United Kingdom, 2007.
18. F. Lawrence, *I Am Legend*, Warner Bros., United States of America, 2007.
19. B. Stoker, *Dracula*, Archibald Constable and Company, London, UK, 1897.
20. R. Preston, *The Hot Zone*, Random House, New York, 1994.
21. M. R. Carey, *The Girl with All the Gifts*, Orbit, Great Britain, 2014.
22. S. Soderbergh, *Contagion*, Warner Bros., United States of America, 2011.
23. M. Grant, *Parasite*, Orbit, Croydon, 2013.
24. R. Matherson, *I Am Legend*, Gold Medal Books, NY, USA, 1956.
25. Urban Dictionary, *Definition: Zoombies*, 2009 [cited 2016 11 13], available from: http://www.urbandictionary.com/define. php?term=Zoombies.
26. E. Wright, *Shaun of the Dead*, United International Pictures, United Kingdom, 2004.
27. J. Levine, *Warm Bodies*, Summit Entertainment, United States of America, 2013.
28. I. Marion, *Warm Bodies*, Vintage, London, 2010.
29. S. G. Browne, *Breathers: A Zombie's Lament*, Broadway Books, US, 2009.
30. H. Martin, executive producer, *In the Flesh*, [Television Broadcast], BBC Drama Productions, UK, 2013.
31. S. Meyer, *Twilight*, Atom, London, UK, 2007.
32. C. Higson, *The Enemy*, Puffin, London, England, 2009.
33. C. Higson, *The Dead*, Puffin, London, England, 2011.
34. C. Higson, *The Fear*, Puffin, London, England, 2011.
35. T. C. Smith, Zombie infections: epidemiology, treatment, and prevention, *Br. Med. J.*, 2015, **351**, h6423.

CHAPTER 5

How to Build a Super Soldier

AKSHAT RATHI

Quartz, London, UK
E-mail: rathi.akshat@gmail.com

5.1 INTRODUCTION

Cartoonists are bound by few rules, especially if their job is to create a superhero. So, it is intriguing that cartoonists Joe Simon and Jack Kirby didn't create a character with infinite power to end the horrifying atrocities of the Nazis. Instead, they created Captain America – a human who performs at his peak and has a strong moral compass.

Their patriotic superhero was an instant success. Even a year before the US joined the Allied Forces, a comic that showed Captain America punching Adolf Hitler sold over a million copies. Although there have been some ups and downs, as there should be in any human life, the legendary character has now spanned more than 70 years of comics, TV and movies. Despite having no superhuman ability, he is ranked sixth on IGN's list of top 100 comic-book heroes of all time.[1]

The Secret Science of Superheroes
Edited by Mark Lorch and Andy Miah
© The Royal Society of Chemistry, 2017
Published by the Royal Society of Chemistry, www.rsc.org

Captain America exposed a long-held fascination of military tacticians to add a new weapon to their arsenal: a super soldier, or ultimate hero. The real-life history of that unfulfilled dream explains why they made Captain America as we know him today. The sole aim of wars, at least in the minds of generals, is to win and this focus has underpinned the development of remarkable innovations in weapons technology: from spears to autonomous drones. But, as the decorated World War II general George Patton put it, 'Wars may be fought with weapons, but they are won by men.' And so the idea of the human victor has remained central to the idea of war. Thus, despite the great scientific progress since Patton's time and the change in the kinds of wars fought, Patton's words remain pertinent even today. Progress in realizing the super soldier, however, has been slow.

It is said that the ancient Spartans chose their fighter's immediately after birth and spared no effort in training them for battle. Michael Hanlon reports that 'Those who failed the first round of selection, which took place at the ripe old age of 48 hours, were left at the foot of a mountain to die'.[2]

The survivors would, in years to come, often wonder if these rejects were the lucky ones. Because to harden them up, putative Spartan warriors were subjected to a vigorous regime involving unending physical violence, severe cold, a lack of sleep and constant sexual abuse.

As wars became bigger, selection became less brutal and, instead, the focus shifted to improving training. The Romans were famous for their ferocious training regimes, which created some of the best fighting armies the world had seen.

Yet, most men in armies were not dying from fighting in wars. Instead, until the dawn of modern science, according to Kendall Hoyt,[3] professor of medicine at Dartmouth College, it was infectious diseases that they could not beat. In the American Civil War, between 1861 and 1865, more than two-thirds of soldiers died because of 'uncontrolled infectious diseases'.[4] In the Spanish-American war of 1898, the proportion of soldiers dying from infections rose to nearly four-fifths. In World War I, nearly half the American casualties in Europe were due to the Spanish flu.

World War II spurred the development of vaccines; as many as 10 vaccines were developed or improved during that period.

These discoveries proved to be the first enhancements that made soldiers invincible against one of nature's deadliest weapons. It is no coincidence that Captain America was created in this era.

5.2   THE STORY OF THE STAR-SPANGLED AVENGER

The biography of Captain America begins as Steven Rogers, who was born in New York City in the 1920s to two Irish immigrants. As a scrawny fine-arts student, he dabbled in writing and drawing comic books. The horrors of the Third Reich motivated him to register for service, but he failed the physical tests. Resolved to serve the nation in some way, he agreed to take part in 'Project: Rebirth,' a secret US military program looking to create enhanced soldiers.

Rogers was given injections of *Super Serum*, a creation of scientist Abraham Erskine, and then exposed to Vita-Rays to trigger enhancement. The process transformed him into Captain America, a human at the zenith of his potential. His powers included enhancements to vision, strength, speed, agility, reflexes, stamina, intelligence and healing. The serum also had other weird effects: an inability to get drunk, an ability to suspend ageing and resistance to hypnosis.

However, immediately after Rogers' 'rebirth', a Nazi spy killed Erskine, though he was shrewd enough in life to know the value of his project to enemies. Consequently, Erskine did not document some crucial elements of the transformation he had created.

There have been many attempts to create super soldiers in the intervening 70 years, though no country has found the perfect formula yet, despite some efforts.

5.3   HOW TO CREATE SUPER POWERS IN MERE MORTALS

5.3.1   Seeing More than Others

When measuring visual acuity, you've probably heard of the term 20/20 score. It's called the Snellen fractions, named after the Dutch ophthalmologist Herman Snellen who developed the scale in 1964. It denotes that you are able to see from a distance of 20 feet what 'normal' people can see from 20 feet. Of course, those with worse vision can achieve a 20/20 score by wearing spectacles.

There are, however, some people who have better than 20/20 vision. Some get a 20/10 score, which means they are able to see from a distance of 20 feet what 'normal' people can only see from 10 feet. If you're not among those few, there are a number of possible ways to achieve superior vision.

An eye is made up of a cornea (which seals the eye), a lens (which focuses rays of lights) and a retina (which converts light to electrical signals which are sent to the brain). Any number of things can give you vision that is worse than those of 'normal' people, but the most common of them occurs when the shape of the eye ball stops the lens from perfectly focusing the rays of light onto the retina. Laser eye surgery, also called LASIK for laser-assisted *in situ* keratomileusis, can fix this problem. The process involves creating a flap on the cornea, then with the help of a computer-assisted laser, cutting out a small amount of the lens. This corrects for the slight aberration in the shape of the eye and can give some people better than 20/20 vision.

Pilot studies done by ophthalmologist Greg Gemoules[5] show that 20/10 vision may also be possible with the use of special lenses. He uses lasers to study the contours in a person's eye and then designs special lenses that improve vision without the need for surgery (though, he warns, it may not work for every one).

But why would you stop at slightly better than normal vision? Why not try to achieve something like 'night vision', which was invented in World War II to use in anti-aircraft defence in the UK. Today, if you have seen any movie about modern warfare, there's no escaping the emerald-green night-vision world, so how about this as a facet of human vision capability?

The benefits of night vision are pretty obvious. Famously, in May 2011, the US SEAL who killed Osama bin Laden was wearing a night-vision goggle. As the *New Yorker*[6] told the story, 'On the top stair, the lead SEAL swivelled right; with his night-vision goggles, he discerned that a tall, rangy man with a fist-length beard was peeking out from behind a bedroom door, ten feet away. The SEAL instantly sensed that it was Crankshaft [codename for bin Laden].'

The technology is simple. Although it is too dark for human eyes to see anything, specialized sensors are able to catch the few photons that are still being reflected from objects. The sensor converts the signal to electrons, then a multiplier fires a set number of extra electrons for each one it detects, and throws them onto a phosphorescent screen. The electrons create tiny flashes of phosphorescence, typically green in colour, and create a world that's easily visible to a human eye.

A more recent invention in enhancing vision is thermal imaging, which captures infrared radiation released as heat. All humans continuously emit heat and would be seen on a thermal-imaging device. Combine all the above and you've given a soldier better vision than Captain America.

5.3.2 Catch Me If You Can

Captain America can run 100 metres in less than half the time it takes Usain Bolt, the Jamaican runner who holds the current world record. What would we need to do to enhance Bolt to beat Rogers?

In 2011, scientists were able to create mighty mice,[7] who ran at twice the speed of normal mice and for 20 minutes longer. Hiroyasu Yamamoto at the Federal Polytechnic School of Lausanne in Switzerland found that mice bred without the NCoR1 gene had those mighty abilities. The reason being that NCoR1 is responsible for making nuclear receptor corepressor 1, which is a protein that acts as a turn-off switch for mitochondria, the powerhouses of every living cell. In the absence of this protein, the mice were able to push the limits of their cells and do more than their 'normal peers.' They even gained more muscle mass without needing more food.

The trouble with this process of knocking out a specific gene is that it can only be used in test animals, whom we can kill if things were to go wrong, and even then there are ethical concerns about doing so. However, there is a way that humans have enhanced their speed without becoming guinea pigs for such brutal scientific experiments and this is through the discovery of new training techniques that allow us to get the most out of what we are born with. A good indication of this can be seen in the last century of Olympic competition. The 100 m sprint times have shown a growing capacity to get more out of our biology by training the body harder and differently. The current men's world record, held by Usain Bolt, shows that he would be more than 10 metres ahead of his winning counterpart from 1900 (Figure 5.1).

That means, over a century, humans have enhanced themselves to run at least 10% faster. These developments have come through tiny, but important, tweaks made to training regime, shoe design, material used to make the running track, and such technological advances that have and continue to help humans to keep breaking that Olympic record. In addition, athletes benefit from enhanced sport science know-how, which has allowed them to push the boundaries of what we thought were the limits of humanity. If we add the uncertainty about how doping technologies are used, then it's possible that even more gains in biological performance could be won, before we begin to even think about genetically modified athletes.

5.3.3 Super Size Me

Outside the Marvel Universe, we may not be able to use the Super Serum and Vita-Rays. However, researchers have discovered how to use hormones to create 'super-soldier ants'[8] and we can learn a lot from them. Ants are known for their highly organized colonies, where labour is divided among workers who build and forage and soldiers who defend. Among the thousand or so species of ants found in the genus *Pheidole*, eight species contains a special caste of giant-headed soldier ants.

While studying a wild *Pheidole morrisi* colony, which wasn't known to have the giant version, Ehab Abouheif of McGill University in Canada found a few ants that were abnormally large. He and his colleagues found that they were actually mutants of their fellow soldier ants. It turned out that they had been created because of hormonal abnormalities during the development stage.

Abouheif reverse-engineered the process and found that a growth hormone called methoprene could be used to create super-soldier ants in *P. morrisi*. Amazingly, he was able to apply the method to at least two other species of *Pheidole* genus not known to have the super-soldier caste. His research indicated that the trait existed in other ants of the genus and could be induced environmentally.

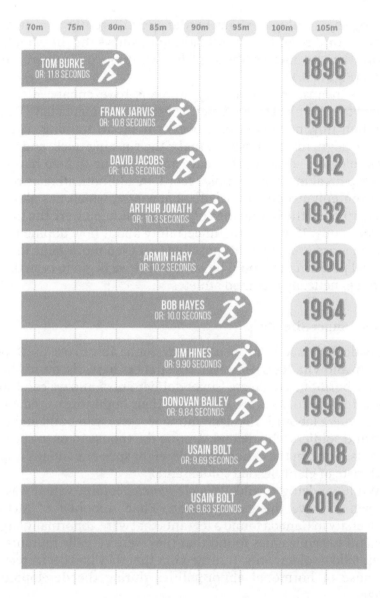

Figure 5.1   100 m times from 1900 to Usain Bolt, compared with Captain America. © Andy Brunning 2017.

190m 195m 200m 205m 210m

CAPTAIN AMERICA
RUNS 212 METRES IN BOLT'S 2012 RECORD TIME!

Perhaps humans also have such a hidden trait that we have yet to discover, but which we could induce using special serums. Until then, we can always rely on drugs and it should be no surprise that soldiers already do.

5.3.4   Keep up with Me

Besides super strength, a super soldier would also benefit from superior stamina. Drawing from existing research, we can estimate that two ingredients[9] of the super-soldier serum were probably erythropoietin (EPO) and hypoxia-inducible factor (HIF). HIF activates EPO which, in turn, stimulates the production of red blood cells. With more red blood cells flowing, the body has a greater capacity to carry oxygen and thus burn more energy, which means we can keep working for longer.

A non-pharmaceutical way to achieve the same outcome is to extract red blood cells from the body and store them for later use. Indeed, this is also a known method of performance enhancement among athletes, known as autologous blood doping.

5.3.5   Show Me the Muscle Power

In his book *Shooting Up*,[10] historian Lukasz Kamienski has charted the history of drug use in warfare. As *Vice* explains, this history is long, beginning with 'Viking berserkers driven into a trance-like frenzy by mushrooms to Inca warriors sustained by coca leaves to American Civil War soldiers hooked on morphine and the speed-fuelled Wehrmacht'.[11] Today, the use continues, but is underpinned by scientific discovery and takes various forms. For instance, Tony Hsia, a former US military officer, wrote in 2013 in the *New York Times*,

> The use of supplements to become bigger, stronger and faster is a fact. When I had to conduct a health and welfare inspection, I was shocked by the profusion and variety of pills and powders I found in the rooms of soldiers. A few rooms could have been confused for mini-pharmacies.[12]

The latest in such artificial substances for enhancements is the use of designer steroids. While there has been a long history of steroid use in sport, designer steroids are even harder to detect and derive from medical research that has yet to reach the commercial market. In the past, such steroid use would come from known pharmaceutical products, which athletes would misuse. Today, athletes are looking deeper into research before it even arrives in pharmacies and are cooking up new drugs. Anabolic steroids masquerade as natural hormones and activate androgenic receptors. This, in turn, increases production of proteins and thus triggers muscle growth, allowing the athletes to build their functional muscle mass. They also cut down on the time required to breakdown waste and thus decrease the time required for recovery, which means an athlete – or a soldier – can train even harder and get an advantage over a competitor or enemy who doesn't use such methods. A 2004 study[13] showed that men who used these steroids for 10 weeks grew between 2 kg and 5 kg of lean mass, that is mostly muscle.

These steroids have a legitimate use in helping patients with, say, cancer to recover. In fact, there are a whole range of muscle-wasting diseases where these substances find medically legitimate uses.

Their application to sports and the military throws up new questions for society to consider. After all, even if it is reasonably clear that an athlete is cheating when using them in sport – where the rules prohibit their use – in the military, no such boundaries exist, so why shouldn't they be used? The arguments against such use are often focused on the fact that these drugs could lead to increased medical risks. In battle, however, if they help a soldier be more likely to stay alive, by allowing them to be stronger than their enemy, this seems like a strong case *for* their use.

In sports, the problem is a little different and it is possible that the enhanced risk to the athlete derives less from the substance and more from the fact that they are used without any proper medical supervision, which could minimize those risks, or ensure the substances are as safe as they can be. Nevertheless, with any synthetic pharmaceutical substance, the side-effects can be harsh. They can cause cardiovascular issues, throw the body's cholesterol balance off, and even increase chances of suffering from depression. In both cases, one would need to consider the short-term gain – in battle or sport – and weigh this up against the long-term harms, but this is always a predicament, particularly in matters of war.

5.3.6   No Shut Eye

Even when we've enhanced the strength and stamina of a soldier, there remain biological restrictions on what they can do. For example, no matter the strength or intelligence of a soldier, each one will have to sleep for about a third of the day, every day. For instance, in Afghanistan, studies suggest that soldiers in combat get only about four hours of rest, which makes sleep deprivation the biggest factor in reduced fighting performance. So, any army that is able to overcome the need for sleep will make huge gains over its enemy.

The US Defense Advanced Research Projects Agency (DARPA) has been at the forefront of such research and is trying to reduce a soldier's need to sleep without compromising their ability to function effectively. Yet, as Chris Berka, the co-founder of Advanced Brain Monitoring (ABM), one of DARPA's research partners, told *Aeon* 'Every so often, a new stimulant comes along, and it works well, and there's a lot of interest, and then you don't hear anything more about it, because it has its limitations.'[14]

Berka has two concrete examples of when this has become apparent. In 2002, the US Air Force was testing the use of amphetamines, popularly called 'go pills.' However, on April 17 in the same year, a top-gun fighter pilot who was flying his jet over Kandahar spotted some soldiers on the ground. The pilot thought he was under attack and killed four soldiers, who later were revealed to be Canadian soldiers performing a drill. The incident resulted in a court martial and the drug's limitations were laid bare for everyone to see.

The other example is that of modafinil, a drug that is prescribed to those suffering from narcolepsy, a condition that can cause sudden, irresistible bouts of sleepiness. Those unaffected by the condition can use the drug to stave off sleep by taking about 50 mg of modafinil every eight hours. Doing so allows someone to focus as if they are a fully rested person, for up to 60 hours. However, there is also a trade-off in doing so, which is something called 'tunnelling.' This is where people make decisions without considering their surroundings and their social situations, just like the pilot on amphetamines who ended up choosing to fire on friendly soldiers.

So, on balance, it seems difficult to argue that modafinil is enhancing a healthy human or, if it is, it is only doing so in a very narrow sense, much narrower than is likely to be helpful in matters of human endeavour, which are often inherently multi-faceted and require a wide range of cognitive capacities.

After decades of trying, researchers have been struggling to beat sleep. However, we still don't know for sure *why* we sleep. It is clear that, since we have evolved to sleep for a third of our life, we can conclude that sleep is a crucial enough function. Indeed, all animals have some form of resting cycle – whether or not we characterize it as sleep. So, it seems that our sleep cycle is something that we would be unwise to modify.

Nevertheless, most recently, researchers have endeavoured to find ways of maximizing the benefits of sleep in the fewest of hours. If we can't do away with sleep, then maybe we can optimize it and reduce the number of hours we need. An experimental technique called transcranial magnetic stimulation (TMS) promises just that. The treatment involves attaching a device to the brain which creates magnetic fields that target areas where slow-wave sleep is generated and then propagate this to the rest of the brain. The promise is that such a device could put a person into deep sleep instantaneously. Full control of our sleep cycle could, in principle, allow us to get the benefits in half as many hours.

5.3.7 By Mere Thought

Another technology like the TMS is called transcranial direct-current stimulation (tDCS). Instead of using magnetic fields, this technique passes tiny shocks of electricity through the brain. In short it is a way of manipulating how your brain's neurons talk without using an invasive surgery. Used properly, tDCS can be like a cognitive enhancer.[15] In DARPA-funded research, soldiers on tDCS learnt certain strategy skills in half the time, compared to those without tDCS. The UK's Defence Science and Research Laboratory has had similar success with tDCS among air force pilots. Even leading tDCS researchers aren't completely sure how such imprecise stimulations can have such precise outcomes. However, it means that, as they get better at precise stimulations, they are more likely to get even better results. As yet, unlike most drugs, tDCS has not revealed significant side-effects and this has given researchers and the military more reason to investigate just how tDCS could be used to make soldiers smarter.

In the *Avengers*, despite having a whole team of superheroes with extraordinary abilities, there is a good reason for why *Captain America* is the leader. He may not be able to smash like Hulk, or summon a storm like Thor, or fly like Iron Man. However, he is able to keep a calm head in any situation and he has the capacity to think and act strategically and, while we haven't yet figured out how to create humans with superior intelligence, there's clearly no stopping us from trying.

Many of the powers that we would need to develop to ensure we have superhuman capacities have biological precedents – we don't need to invent then, we can simply look to nature and see if we can emulate them. Whether it is heat-seeking vision – enjoyed by snakes – or the capacity to regenerate limbs – as in the case of the salamander – or the production of the extraordinarily strong spider silk – we may find that becoming superhuman in the future will mean looking to nature to discover what can take us to the next stage of human evolution.

REFERENCES

1. *IGN's Top 100 Comic Book Heroes*, 2016 [cited 8 November 2016], available from: http://uk.ign.com/top/comic-book-heroes/6.
2. M. Hanlon, *'Super Soldiers': The Quest for the Ultimate Human Killing Machine*, The Independent, 2011 [cited 8 February 2007], available from: http://www.independent.co.uk/news/science/super-soldiers-the-quest-for-the-ultimate-human-killing-machine-6263279.html.
3. K. Hoyt, *How World War II Spurred Vaccine Innovation*, The Conversation, 2015 [cited 8 November 2016], available from: https://theconversation.com/how-world-war-ii-spurred-vaccine-innovation-39903.
4. J. Sartin, Infectious diseases during the Civil War: The triumph of the "Third Army", *Clin. Infect. Dis.*, 1993, **16**(4), 580–584.

5. G. Gemoules, *20/10 Vision Explained and Achievable*, 2015, available from: https://laserfitlens.com/laserfit-scleral-lenses-2010-vision-who-can-benefit/.

6. N. Schmidle, *Getting Bin Laden*, The New Yorker, 2011 [cited 8 November 2016], available from: http://www.newyorker.com/magazine/2011/08/08/getting-bin-laden.

7. H. Yamamoto, E. Williams, L. Mouchiroud, C. Cantó, W. Fan and M. Downes, *et al.*, NCoR1 Is a conserved physiological modulator of muscle mass and oxidative function, *Cell*, 2011, **147**(4), 827–839.

8. R. Rajakumar, D. San Mauro, M. Dijkstra, M. Huang, D. Wheeler and F. Hiou-Tim, *et al.*, Ancestral developmental potential facilitates parallel evolution in ants, *Science*, 2012, **335**(6064), 79–82.

9. J. Bryner, *What is Blood Doping?* Live Science, 2017 [cited 8 November 2016], available from: http://www.livescience.com/32388-what-is-blood-doping.html.

10. L. Kamienski, *Shooting Up: A History of Drugs in Warfare*, C. Hurst & Co., 2016.

11. O. Rickett, *Soldiers Have Used Drugs to Enhance Their Killing Capabilities in Basically Every War*, Vice, 2016 [cited 8 November 2016], available from: https://www.vice.com/en_us/article/drugs-have-been-used-in-pretty-much-every-war-ever-shooting-up.

12. T. Hsia, *The Performance-Enhanced Military*, [Internet], At War Blog, 2017 [cited 8 November 2016], available from: http://atwar.blogs.nytimes.com/2010/05/07/the-performance-enhanced-military.

13. F. Hartgens and H. Kuipers, Effects of androgenic-anabolic steroids in athletes, *Sports Med.*, 2004, **34**(8), 513–554.

14. B. Hains, *Technology to Cut down on Sleep is Just Around the Corner–Jessa Gamble | Aeon Essays*, Aeon, 2013 [cited 8 November 2016], available from: https://aeon.co/essays/technology-to-cut-down-on-sleep-is-just-around-the-corner.

15. E. Young, *Can You Supercharge Your Brain?* Mosaic, 2014 [cited 8 November 2016], available from: https://mosaic-science.com/story/can-you-supercharge-your-brain.

CHAPTER 6

The Real World Super Metal

PAUL R. COXON

Department of Materials Science & Metallurgy, University of
Cambridge, 27 Charles Babbage Road, Cambridge, CB3 0FS, UK
E-mail: prc39@cam.ac.uk

6.1 INTRODUCTION

Other chapters in this book spend a good deal of time speculating about the nature of comic book supermaterials. They look at the fictional vibranium, imagine how graphene might be used within supersuits, theorize on how research on memory materials could lead to such creations as Batman's wing suits, and discuss meta-materials that might make the wearer invisible. My fellow authors do a sterling job of linking the science fiction of superheroes with the science fact that exists in labs around the world. Maybe, in the not too distant future, that lab-based science will morph into technologies not dissimilar to those of the comic books. However, amongst all the speculation on what might be possible in the future, perhaps we've overlooked materials that are already 'super'; elements without which much of the modern world's technologies would not exist. Many of these

The Secret Science of Superheroes
Edited by Mark Lorch and Andy Miah
© The Royal Society of Chemistry, 2017
Published by the Royal Society of Chemistry, www.rsc.org

elements are also in short supply, and so we could really do with a superhero to deploy their powers and manufacture some more. This chapter explores one specific case of these materials, lithium, to show just how amazing our world already is, before we even get into science fiction.

Let's start by perusing the periodic table and checking out those vital but endangered elements (Figure 6.1). Somewhat surprisingly, the first element in short supply is helium. Given that it is the second most abundant element in the universe and turns up in party balloons you may wonder what the problem is. Basically, there's masses of it out there (mainly in stars), but precious few terrestrial sources. Not only that, but once it gets into the atmosphere it just floats off into space. You may wonder what the fuss is about, after all civilization is unlikely to collapse for want of party balloons, but helium has a host of hi-tech applications – from medical imaging, to super particle colliders, lasers and gas mixtures used by divers.

Then there is a whole host of elements smack in the middle of the table, such as indium – a vital component of touchscreen devices. There is hafnium, which has remarkable properties, including a super high melting point (2506 K) and an ability to absorb neutrons – making it hugely useful in controlling nuclear reactions. And when mixed with other metals hafnium makes an alloy that is super resistant to heat and oxidation; perfect for rocket engines. Meanwhile, many of the other elements perform roles as catalysts, speeding up chemical reactions in industry but also in your car's catalytic converters.

However, there is one element that stands out from all those others, nestled below hydrogen and is characterized by the abbreviation 'Li' — that's lithium, slightly overlooked by other elements. In recent years, this innocuous element has been plucked from relative obscurity to superpower our modern world and has the potential to revolutionize transport, renewable energy and mobile devices. Its role in the world was foreseen by the authors of the Dr Manhattan graphic novel back in 1987.[1] In a single comic strip panel in the original *Watchmen* series, there is a casual remark which says that, by the late 1980s, fossil fuels cars are obsolete. In the comic book world electric vehicles are everywhere, all thanks to Doctor M's ability to synthesize lithium and mass-produce lithium batteries.

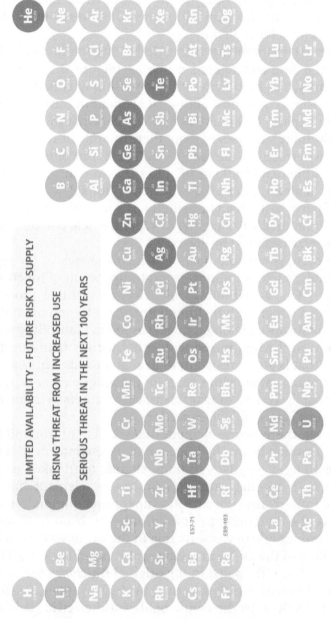

Figure 6.1 The periodic table, coded by availability. © Andy Brunning 2017.

6.2 JON OSTERMAN: WHERE IS HE FROM AND AN ACCIDENT

Doctor Manhattan is one of the most recognizable characters in the DC universe. He first appeared in the *Watchmen* series and as his story develops, he gradually transforms from humble lab technician into perhaps the most omnipotent of all superheroes.[1]

Jon attends Princeton University and graduates with a PhD in atomic physics and takes a research position in a laboratory where he takes part in experiments exploring the 'intrinsic fields' of physical objects which, if tampered with, result in their disintegration. Soon after starting here, an accident occurs in which Jon is caught in the fan experimental field chamber. His physical body is vaporized and he is officially declared dead. However, Jon's consciousness survives and, over the following months, he gradually reappears before assuming his most famous form, a tall and hairless blue-skinned man – 'a quantum being of unlimited power'. After this transformation, the now god-like Doctor Manhattan becomes aware of profound changes in his senses, intellect and abilities. He possesses truly superhuman powers; he is immortal, he can fly, rearrange matter outside the normal laws of physics using only his mind, he can manipulate his size and density, and be in more than one place simultaneously. He experiences time in a non-linear fashion and can view all history at once. As his story develops, Doctor Manhattan's powers revolutionize the world, and radically alter the world's economy.

Indeed, in real-life, the ability to harness and produce metals and materials economically has been one of humanity's longest pursuits. Since Doctor Manhattan has control over quantum fields and can easily create any element at will, this energy revolution is but a cheap party trick. At the time, the idea of electric vehicles was still just getting off the ground – and gas-guzzlers still reigned supreme. By the early 21st century, however, electric vehicles powered by lithium batteries are commonplace and are becoming more popular. So, let's take a closer look at how this supermaterial functions in our world.

6.3 LITHIUM: A HISTORY

Without our very own Dr Manhattan, how do we create lithium today? We can't, but thankfully the universe did that job for us over 14 billion of years ago. It's quite humbling to think that when we hold a mobile phone or tablet in our hands, we are also grasping one of the oldest materials in the universe. In recent years, the periodic table has expanded due to the arrival of an increasing number of new and synthetic elements forged in the heart of industrial-scale particle colliders. In comparison, lithium is a veritable Methuselah; it was formed before the stars and galaxies themselves. Lithium was born in the first chaotic minutes after the Big Bang, at a time just after the first subatomic particles took shape. By the time the cosmos was 10 seconds old it had cooled to a mere 2 billion degrees centigrade and these swirling protons and neutrons began to combine and form the first atomic nuclei. This process is called primordial nucleosynthesis – and it made our very first elements: hydrogen, helium and a tiny amount of lithium. Lithium is the third element in the periodic table and, as such, is the lightest solid at room temperature.

Elements come in a variety of forms, known as isotopes, which differ by the number of neutrons they have in their nucleus. Lithium always has three protons and, in nature, it comes in two isotopes Li-6 and Li-7, each with three protons and either three or four neutrons in their nuclei respectively. Over 90% of lithium is in the form of Li-7. So, when thinking about what it might mean to have a power like Dr Manhattan's – the power to create any kind of material – we need to come to terms with the fact that some materials have a unique and long history, which presents a, perhaps, insurmountable obstacle to our actually designing such materials from scratch today. In other words, the existence of a superpower may compromise some historically determined processes, which makes it impossible to imagine in our real world.

While some materials become less effective as they get older, lithium belies its great age and maintains the bounce and vigour of a two-year-old toddler. As a metal, lithium is an excellent conductor of heat and electricity and, in its elemental form, is extremely reactive in air and moisture – so much so that special measures must be taken to store it safely. You have probably seen the popular classroom demonstration of dropping a small lump of lithium in a tray of water and watching it gently swirl and skip over the surface in a wildly fizzing ball. Lithium has the lowest density of all metals with a density of 0.53 g cm^{-3}. This is about half that of water, meaning that lithium actually floats, and can react with the water to give off hydrogen gas.[2] Pure lithium on its own is a silvery-white metal, soft enough for it to be sliced like a piece of cheese. When freshly cut, the shiny metallic lustre can be seen, but the surface soon tarnishes as it reacts with nitrogen in the atmosphere to form black lithium nitride. This colour change is an impressive chemical feat in itself. Getting a material to do pretty much anything with nitrogen gas is a real challenge. Given how much nitrogen there is surrounding us all the time one would think that it should be pretty easy to obtain, however nitrogen atoms form robust triple bonds between each other making it very difficult to break up and react with anything else.

As befits such a reactive element, it is unsurprising that the story of how lithium was discovered is so flamboyant. Since pure lithium is too reactive to exist on its own, it is normally locked up, safely bound in a chemical prison as a mineral with other elements. One of these lithium-containing minerals, petalite, was discovered by the Brazilian naturalist and statesman Jozé Bonifácio de Andrada during a tour of Scandinavia in the last years of the 18th century. Bonifácio is described as 'the Greatest Man in Brazilian History', whose life bestrides politics and science. After a glittering scientific career, he entered politics and became First Minister of Brazil (1822–23) and a strong advocate for Brazilian independence. A highly educated man, using political guile and savvy, he managed to keep the country united while other nations on the continent spilt apart. He is also known throughout Brazil as the nation's Founding Father, with over 100 books written about him in Portuguese, though he's largely unknown by the English-speaking world.

Born in Sao Paulo in 1763, Bonifácio arrived from Brazil at the Royal School of Mines in Paris in 1790 to study chemistry, mineralogy and botany with the leading scientists of the age. From 1792 to 1798, after a spell in Saxony at the world-famous Freiburg University of Mining and Technology, he travelled throughout Europe – visiting the mines in France, Italy, Hungary, the Netherlands and Scandinavia. All the while he was expanding his knowledge of rocks and how they formed. Mineralogy and geology was a hot topic at this time since mineral wealth helped drive the nations' economies and provided the raw materials behind the Industrial Revolution.

It was during his visit in Sweden on the island of Utö that Bonifácio found a new and interesting mineral. Sweden has an interesting geology that makes it a treasure trove of new minerals. Four elements of the periodic table were discovered in a single quarry nearby. Utö is an island in the Stockholm archipelago where a small silver mine opened in 1607. It was soon closed, as the amount of silver it produced was too small, but the mining of iron ore – a hematite with an iron content of 40% – began there in the Middle Ages and expanded in the beginning of the 17th century. The ore was taken for further processing on the Swedish mainland.

In the biggest and deepest mine special mineral veins were found which were rich in many non-ferrous minerals, especially petalite. As far as Bonifácio knew this new mineral had not been described earlier. He called it petalite and we now know it is a compound of lithium, aluminium and silicon ($LiAISi_4O_{10}$). Petalite was observed to give an intense crimson flame when thrown onto a fire. The difficulty was in isolating the lithium and an attempt at this challenge was met 17 years later in 1817 by Johan August Arfwedson, a young Swedish chemist working in the laboratory of Jöns Jacob Berzelius in Stockholm.

Arfwedson analysed the mineral and deduced that it contained a previously unknown element, a new alkali metal that was lighter than sodium. Arfwedson managed to break the petalite down into a lithium salt, no mean feat, and although he was never able to fully isolate it in its pure form, he is today credited as the discoverer of lithium. Berzelius named the new mineral, 'lithos,' from the Greek word for 'stone.'

Soon after, the English chemist William Thomas Brande isolated lithium in 1821 by passing a current through lithium oxide (Li_2O). Soon lithium was found in other minerals, in sea water and in the waters of fashionable spas.

Some of the very earliest uses of lithium were in medicine. In the mid-19th century it was found that a solution of lithium carbonate could help treat people with kidney stones and gout, by dissolving the hard and painful urate crystals. Because these conditions came to be associated with 'melancholy' and 'mania', lithium salts were soon used as tranquillizers, although this may seem strange given lithium's notoriously volatile behaviour. The first use of lithium carbonate to prevent depression came in 1886 and lithium pharmaceuticals are still some of the most effective therapies we have for the treatment of psychological conditions today.[3]

Despite its potency, lithium is not needed in our bodies – unlike iron, magnesium or phosphorus which are absolutely vital. But although lithium isn't essential for biological functions its effect on the nervous system is profound and can help people with mental illness return to normal life. The exact reasons for this are still not known for certain. Lithium can increase the production of serotonin, a chemical that is associated with euphoria, and it has an influence on neurotransmitters and nerve signalling, most likely due to its chemical similarity to sodium. There is some evidence to suggest that communities whose water supplies contain trace amounts of lithium mineral have much lower suicide rates.[4]

Today lithium has many applications and is deemed a critical economic element. Only a small fraction is used in pharmaceuticals. Around half of the 175 000 tonnes of lithium minerals produced each year goes into the production of special ceramics and glasses. Magnesium-lithium alloys can be used for armour plating in military vehicles. Lithium fluoride has one of the lowest refractive indexes and its crystals are used in specialist optical applications. Lithium also finds use in heat-transfer applications and as a base for car/cycle lubricants. As the lightest metal, a common use of lithium is in the production of lightweight alloys for aircraft. Lighter materials help reduce fuel costs, where each kilogram reduction can save 125 litres of fuel per year.[5]

Of all lithium's uses, however, the one which has seen the largest growth and the one with the biggest future implications is in energy storage. Over the space of 30 years lithium has become synonymous with batteries. There are many different kinds of lithium batteries and they have made a large impact on how we live our lives today. Rechargeable Li-ion batteries have helped drive the modern digital era – they powered the mobile phones of the early 1990s and are inside the laptop computers, digital cameras, MP3 players, smartphones and tablets of today. When you need a small and lightweight battery, lithium is usually involved. The lithium battery market is growing at a tremendous rate and is poised to get even bigger over the coming century as lithium becomes a crucial component of electric/hybrid vehicle batteries and long-term energy storage.

As our world turns away from fossil fuels and towards renewable energy sources, advanced batteries will play an even greater role in our lives. Solar photovoltaics now provide 1–1.5% of the world's electricity and are expected to become one of the dominant sources of electricity this century as manufacturing costs fall. Unfortunately, they are by their very nature intermittent. They don't generate power at night, which is when we often have the highest energy demand in our homes. Storing this renewable energy should help maintain grid stability and lessen the issues of pollution and global warming by reducing our dependence on fossil fuels. As the energy demands of the world become greater, energy storage will become ever more important, and Li-ion batteries are poised as the technology of choice to enable the mass market take-up of renewable energy technology and electric vehicles.

6.4   WHAT ARE LITHIUM BATTERIES? HOW DO THEY WORK?

Lithium's high reactivity is largely down to its atoms and their eagerness to be rid of their outer electron. This also makes lithium an excellent battery material. A battery can be thought of as a frustrated, high-energy chemical bomb, where all the chemical energy is locked up between the battery electrode materials. Placing an electrolyte between the two electrodes stops all the energy being released at once, creating a chemical bridge to allow the energy to be released in a controlled and safe manner as and when we need it. A typical positive electrode is made from cobalt or manganese oxide and the negative electrode is usually made from graphite, a layered form of carbon. The electrolyte (a liquid surrounding the electrodes) is usually made from lithium salts dissolved in an organic solvent. For a rechargeable Li-ion battery, each lithium atom can shed one electron:

$$Li \rightarrow Li^+ + e^-$$

As the battery discharges, one lithium atom at the negative electrode splits into a lithium ion and an electron; current is drawn as the lithium ion travels through the electrolyte, and the electron exits the battery and flows around the external circuit. As the electron flows through the circuit and back into the battery *via* the positive electrode, it meets the lithium ion and recombines to form lithium. During recharging, the reverse process takes place.[6]

The key advantage of using lithium is because it is a small and light element, which means its ions can easily slide between the layers of the electrodes. This makes lithium the most energy dense of battery materials – meaning it stores the most energy for a given weight. This allows for very light batteries in applications where the energy density per unit weight is important – such as electric vehicles. In its pure form lithium has the same energy density as petroleum, but in practice the lithium is mixed with other materials that reduce the energy density.

Today there is considerable research all over the world into improving the energy density and lifetimes of lithium-based batteries. New chemistries based upon lithium-sulfur or lithium-air batteries for electric vehicle batteries are soon to become the largest commercial application of lithium and there is real concern supplies may run out in the next 50 years. Without our own Doctor Manhattan, we have to rely on natural lithium sources around the world.

6.5   HOW DO WE GET LITHIUM TODAY?
TOP LITHIUM-PRODUCING COUNTRIES

Lithium is found around the world in rocks and in brine lake deposits that contain lithium chloride. There are over 100 minerals which contain lithium, but the one which contains the most is spodumene. Extracting lithium from hard rocks is laborious and expensive, and so the majority of the world's lithium (>80%) today comes from brine lakes and salt pans. These are large lakes where the salty water is first pumped out of the lake into a series of shallow ponds. Left in the open sun, the water evaporates leaving behind a lithium chloride brine, which is subsequently treated with soda to precipitate and form a white powder of lithium carbonate Li_2CO_3.[7] There is a lot of lithium in sea water, but its extraction is difficult and expensive.

Over half of the world's lithium comes from a large high-altitude area of lakes and brilliant white salt flats in South America which straddles the borders of Bolivia, Chile and Argentina and is known as the 'Lithium Triangle'. Owing to the rapidly increasing demand for lithium and its status as a critical battery material, these large natural deposits of this 'white petroleum' have sparked a rush of investment in this barren and bleak part of the world. It is hard to predict the stability of supply of this vital material as it becomes even crucial in the future. Yet, one thing is clear. This extraordinary material is every bit as amazing as anything imagined in superhero stories and, without it, a great deal of our modern life would not be possible.

REFERENCES

1. A. Moore and D. Gibbons, *Watchmen*, DC Comics Inc., New York, 1st edn, 1987.
2. *Lithium–Element Information, Properties and Uses*, Periodic Table, 2016 [cited 9 November 2016], available from: http://www.rsc.org/periodic-table/element/3/lithium.
3. R. Howland, *The History of Lithium*, PsychEducation, Psycheducation.org, 2007 [cited 9 November 2016], available from: http://psycheducation.org/treatment/mood-stabilizers/the-big-three-for-bipolar-depression/lithium/the-history-of-lithium.
4. H. Ohgami, T. Terao, I. Shiotsuki, N. Ishii and N. Iwata, Lithium levels in drinking water and risk of suicide, *Br. J. Psychiatry*, 2009, **194**(5), 464–465.
5. J. Emsley, *Nature's Building Blocks*, Oxford University Press, Oxford, 1st edn, 2011.
6. E. Eason, *World Lithium Supply*, Large.stanford.edu, 2010 [cited 9 November 2016], available from: http://large.stanford.edu/courses/2010/ph240/eason2/.
7. J. Tarascon, Is lithium the new gold? *Nat. Chem.*, 2010, **2**(6), 510.

CHAPTER 7

Is It a Ceramic? Is It Graphene? No It's Vibranium!

MARK J. WHITING

Department of Mechanical Engineering Sciences, University of Surrey, Guildford, GU2 7XH, UK
E-mail: m.whiting@surrey.ac.uk

7.1 SUPERMATERIALS

Each of the chapters in this book recount the feats of super-heroes. We have records of the fastest, the strongest, the most elastic, the most glutinous and so on, and we have unpacked how their abilities might be possible in our physical 'real' world. Superheroes are all about extremes. As they engage our imaginations they push the boundaries of the possible, but what of the objects and materials that go beyond the ordinary? Are regenerating bones, indestructible shields and impenetrable armour just tools for storytelling, or might they obey the laws of physics and the science of advanced materials?

One of the best-known artefacts in the multiplicity of superhero universes is Captain America's shield. This iconic object is made

The Secret Science of Superheroes
Edited by Mark Lorch and Andy Miah
© The Royal Society of Chemistry, 2017
Published by the Royal Society of Chemistry, www.rsc.org

from vibranium – the supermaterial which we will explore in this chapter. Our journey starts with some scientific observations of what vibranium can do, as well as some claims made about it in the Marvel Cinematic Universe.[†] Hopefully, from these observations we'll be able to figure out some of vibranium's properties. This journey will enable us to propose a working hypothesis about the nature of vibranium, which can be tested, as further films and TV series provide further data.

7.2   OBSERVING VIBRANIUM

Careful observation lies at the heart of science, so what can we observe about vibranium? In the Marvel Cinematic Universe it is present in a number of films as the material from which Captain America's shield is made. The first movie appearance of Captain America and his shield, *Captain America: The First Avenger* (2011), provides us with some hard data to work with. The shield is made by Howard Stark, who claims that it is stronger than steel. We will explore what this actually means a little later in this chapter. It is also said to be only one third of the weight of steel – more precisely, Howard Stark probably means it is a third of the density of steel.

Vibranium's name, we learn, comes from the fact that it is 'completely vibration absorbent', to quote Howard Stark. In the same film Stark also alludes to its extreme rarity. He claims that all of the United States of America's vibranium has been used to make the shield. All of these data are from the time of the Second World War and some caution might be needed as Howard Stark is prone to playing a little fast and loose with the truth. However, later films clarify that vibranium is very rare. For instance, in *Avengers: Age of Ultron* (2015), which is set in the first quarter of the twenty-first century, we discover that vibranium is available, but at immense cost. In this film, vibranium plays a key role as the eponymous Ultron buys vibranium from the South African arms dealer Ulysses Klaue, shortly before removing one of his arms in a pique of temper – the moral of the story being, never trust a sentient AI megalomaniac

---

[†]The story of vibranium in Marvel Comics is a little different to that in the Marvel Cinematic Universe. For simplicity this chapter will only consider the Marvel Cinematic Universe. Anyone wanting to pursue the alternative accounts of vibranium should refer to *Daredevil*, volume 1, #13, February 1966 and *Captain America*, volume 1, #303 and #304, March and April 1985.

robot. This vibranium is destined to provide armour for Ultron's new body and to build a weapon of mass destruction which will turn a part of the nation of Sokovia into a planetary extinction event.

In *Captain America: Civil War* (2016) vibranium is used by T'Challa to make both his Black Panther body armour and his retractable claws. In the same film we discover that T'Challa hails from the African Kingdom of Wakanda, which is apparently the only source of vibranium. The escapades of Captain America and the more recent activities of Black Panther provide valuable data about the key properties and characteristics we have to work with, namely vibranium's strength, density, vibration absorption capability and rarity. These four characteristics will help us answer the question: 'what is vibranium?'. However, before we do this, we can gain further information to clarify the magnitude of its strength and other mechanical properties.

Much of what we observe as Captain America uses his shield for both defence and offence supports the claims of Howard Stark. Its strength is truly formidable – there are numerous occasions when Captain America's shield defeats various projectiles. The shield not only stops bullets from a handgun fired at close range by a petulant Peggy Carter and those from heavy calibre machine guns, but it even survives rocket grenades.

Such behaviour reveals that vibranium is not only amazingly strong, but that it is also tough and hard. Clarity is needed concerning the distinction between these properties. Common English language usage does little to help us, as we tend to use terms like strength, toughness and hardness interchangeably. For example, we can talk about a person having a strong personality, possessing a tough character or being 'as hard as nails', and such terms are essentially synonymous in everyday usage. This is *not* the case when used technically. In materials science, 'strength' is a measure of how easy it is to deform something, 'toughness' is about the energy required to crack a material and 'hardness' is a material's resistance to indentation.

From what we see in the films, vibranium is not only incredibly strong, but also tough and hard. For example, it can block heavily armoured doors while they are closing and, in the process, leave 2 cm or so deep cuts in the plate steel, remaining undamaged in the process (*Captain America: The First Avenger*, 2011), providing evidence that it is much stronger and harder than steel.

The full extent of vibranium's energy-absorbing capability is seen in two later films. In *The Avengers* (2012), Thor confronts Iron Man and Captain America before they appreciate that they share some common objectives. During their melee Captain America uses his shield to ward off a blow from Thor's hammer, Mjölnir. Remarkably, the vibranium shield is able to absorb the energy of this impact as well as reflect some of the energy; the result of the impact of hammer on shield knocks Thor off his feet. In *Captain America: The Winter Soldier* (2014), Captain America jumps from a height of many storeys and uses his shield to take the impact as he hits the ground. The shield absorbs the energy, surviving unscathed and protecting the Captain too. Any explanation of vibranium must account for this remarkable vibration and energy-absorbing capability – this is after all what gives vibranium its name.

Figure 7.1 contains information on vibranium and a range of other materials. We will refer to the data in this infographic throughout this chapter.

7.3   STRONGER THAN STEEL

If vibranium is stronger than steel, then it will help us get closer to its characteristics if we know the properties of the strongest steel. Firstly, though, we need to understand what is meant by 'strength' in a little more detail.

There are many different ways to measure and describe a material's strength. One key consideration is how a force is applied to the material in question. If we pull something apart, then we call this tension and this leads to the concept of a tensile strength. In this vein the ultimate tensile strength is the stress (force applied divided by cross-sectional area) required to break a material – this will usually result in a fracture and the creation of two pieces of material. Measuring the stress to fracture a material is one way to define strength.[‡]

[‡]How the shield copes with repeated stress is also important. When a material survives an extreme load there is no guarantee that it will survive the same load next time. This is because materials can be damaged by loading, and repeated loading can lead to the accumulation of damage to the point where they fail after a certain number of repeated loading events (known as cycles). This consideration leads to the concepts of creep strength and fatigue strength. Fatigue strength is relevant too. However, we've got enough to get through already so I'm afraid we won't be exploring either creep strength or fatigue strength.

ALUMINA
Density (kg per metre cubed) 3900
Young's Modulus (gigapascals) 350
Strength (compressive fracture) 2200
Rarity (arbitrary ranking) 7TH
BORON CARBIDE
Density (kg per metre cubed) 2500
Young's Modulus (gigapascals) 450
Strength (compressive fracture) 5500
Rarity (arbitrary ranking) 4TH
DIAMOND
Density (kg per metre cubed) 3500
Young's Modulus (gigapascals) 1200
Strength (compressive fracture) 20000
Rarity (arbitrary ranking) 2ND
MARAGING STEEL
Density (kg per metre cubed) 8100
Young's Modulus (gigapascals) 210
Strength (tensile yield) 2000
Rarity (arbitrary ranking) 8TH
Ti-6Al-4V
Density (kg per metre cubed) 3900
Young's Modulus (gigapascals) 350
Strength (compressive fracture) 2200
Rarity (arbitrary ranking) 7TH
CARBON FIBRE COMPOSITES
Density (kg per metre cubed) 1600
Young's Modulus (gigapascals) 180
Strength (tensile fracture) 800
Rarity (arbitrary ranking) 5TH
GRAPHENE
Density (kg per metre cubed) 2200
Young's Modulus (gigapascals) 1000
Strength (intrinsic, estimated) 130000
Rarity (arbitrary ranking) 3RD
VIBRANIUM
Density (kg per metre cubed) 2700
Young's Modulus (gigapascals) ?
Strength ?
Rarity (arbitrary ranking) 1ST

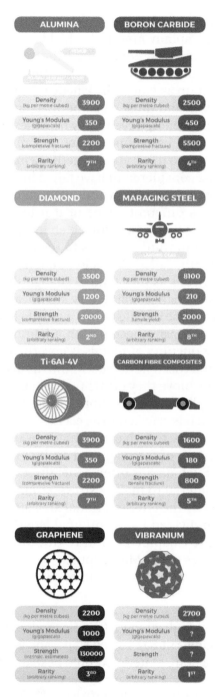

Figure 7.1 Vibranium compared with all sorts of other materials. Values of strength are in MPa. © Andy Brunning 2017.

Measures of strength also need to account for temperature. This is important for Captain America's shield, as it will experience elevated temperature when defeating heat weapons such as that used by Hydra agents in *Captain America: The First Avenger* (2011). It might even heat up as it absorbs kinetic energy and vibrations – energy is conserved and must go somewhere and thermodynamics tells us that most energy eventually becomes thermal energy.

Defining a material based on its fracture strength alone probably isn't the best way to think about whether a material is 'super'. A material would be considered to have failed in many applications if it were to change shape significantly. If Captain America's shield ended up having a different shape or if it lost its aerodynamic properties by losing its slightly domed shape, then this would be a failure. This sort of thinking leads to the definition of another concept for strength, called the yield strength. This is especially appropriate for metals as they tend to change shape prior to fracture. Some metals change shape very significantly prior to fracture and we term these ductile (when a tensile load is applied) or malleable (when they are beaten with a hammer or squashed). The origin of the ductility of metals is governed by the existence and mobility of microscopic defects known as dislocations. When these dislocations move they create permanent changes in shape which is termed plastic deformation – as opposed to recoverable deformation, which is referred to as elastic.[§]

Some measures of strength can be made by using tension tests and plotting a graph of force applied against increase in length of the material. Such data is often generalized by accounting for the size of the material to make a stress–strain plot; stress being the force divided by the cross-sectional area and strain being the change in length divided by the original length (as described in more detail in the Chapter 11, You've Got to Learn to Be More Flexible: Mechanics of the Marvellous).

Figure 7.2 shows the stress–strain response of five very different materials.

- Material A has high strength (the plot ends at high stress), high Young's modulus (it has a steep slope) and is brittle (there is no non-linear region). Material A could be diamond or alumina.

[§]For a more complete account of elasticity see David Jesson's account of Elastigirl's stretchiness in Chapter 11.

- Material B has low strength, moderate stiffness and is also brittle. This material could be silica glass, like you'd find in your windows.
- Material C has a high ultimate tensile strength (maximum stress) and its yield strength is about half this value (the stress where yield, *i.e.* plastic deformation, commences). It also has moderate stiffness and has some ductility (*i.e.* a plastic non-linear region on the curve). This material could be maraging steel, often used in golf clubs and aircraft landing gear.
- Material D has both a lower ultimate tensile strength and a lower yield strength than Material C. It has a similar modulus and much greater ductility than Material C. Material D could be a titanium alloy.
- Material E is a hypothetical curve for vibranium; we will return to this later as it requires further explanation.

Observations of Captain America's shield indicate that it is metallic. In *Captain America: The First Avenger* (2011) we see the shield in its pristine, *i.e.* uncoloured, state. It is seen to exhibit a metallic lustre. Further evidence that vibranium is metallic arises from subsequent observations of the shield in its finished state, when it is emblazoned with the familiar star and the various colours which refer to those of the USA's flag.

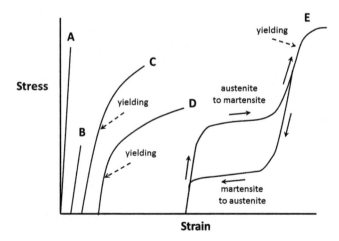

Figure 7.2   The stress–strain response of five materials.

The finished shield still has a lustre which indicates its metallic nature, but the colouration is consistent with it having been anodized.

Anodizing is a process whereby the very thin (c. 4 nm) passive native oxide[¶] formed on the surface of some metals can be thickened (to a few hundred nm). Oxides form when the metal reacts with oxygen in the air, which is especially important for both aluminium and titanium where the layer of metal oxides can be both protective and decorative. The shield's coloured metallic lustre indicates that, like these two metals, vibranium can form a passive oxide which can be deliberately thickened and also coloured.

Having accepted that vibranium is a metal, we also expect that it will fail by yielding rather than by brittle fracture.[||] A significant change in shape would render Captain America's shield less functional and, as we are assuming that it is metallic, then the measure of strength used in this chapter will be yield strength when we refer to vibranium and other metals.**

The infographic (Figure 7.1) compares the strengths of various materials, in MPa. The materials chosen for comparison with vibranium are among the highest performance materials in terms of strength. Carbon fibre composites, titanium alloys and engineering steels are amongst the strongest materials when it comes to materials that can withstand tensile forces. As we have seen in our observations, vibranium is said to be stronger than steel. The 2000 MPa strength for maraging steel would have been the sort of reference point familiar to Howard Stark in the 1940s. Today there are steels that can reach around 7–10 000 MPa in strength, according to Hammond *et al.*[1] These are known as super bainite steels. These steels are used in advanced armour applications, but their cost means that they have yet to find application in more everyday roles.

¶The term passive oxide is a technical one that refers to an oxide which is thin to the point of transparency, highly adherent and protective of the underlying metal.

||We do see what looks like a brittle fracture of the Captain's shield in '*Avengers: Age of Ultron*'; however, the scene takes place whilst Tony Stark is hallucinating, so this probably isn't a reliable source.

**This term is sometimes given different names, such as 'the limit of proportionality', 'the elastic limit' and engineers use something similar called a proof stress.

7.4   AS DENSE AS ALUMINIUM

Does this mean that vibranium could plausibly be an aptly-named super bainite material, ahead of its time? Well, there are several reasons why this conclusion does not fit the data we have already obtained from our observations of the Marvel Cinematic Universe. The most obvious 'deal breaker' is vibranium's density – all steels have very similar densities and bainitic steels are no exception.

If vibranium has one-third the density of steel this means it is close to that of aluminium at 2700 kg m^{-3}. However, even the strongest aluminium alloys are pitifully weak compared to the strongest steels. The strongest commercial aluminium alloys have strengths of around 600 MPa.[††] Aluminium alloys are also very soft despite their reasonable strength. Although the anodizing we mentioned earlier will cause a slight improvement in hardness, an ordinary kitchen knife would scratch the Captain's shield if it were made from aluminium. Our observations of the Marvel Cinematic Universe reveal that conventional materials leave it unscathed – only the vibranium of Black Panther's claws have thus far left a clear mark on it (*Captain America: Civil War*, 2016).

If vibranium is significantly stronger than steel and has the density of aluminium, then this indicates that it is not simply an alloy, based on the strongest known alloy systems of today. Other very strong alloys that have emerged over the past few decades do not fit these properties. For instance, both nickel-based superalloys, which are used in the hottest part of jet engines, and cobalt–chromium alloys, which make the strongest biomedical implant materials, are of similar strength to maraging steel. They are nowhere near the strength of super bainite steels. They both also have very a similar density to steels and thus fall foul of our 'one-third the density of steel' criteria.

7.5   ELEMENTS: NEW AND OLD

Does this mean that vibranium is a hitherto unknown element? New elements have been a staple feature of superhero stories in both comics and on screen. However, there are some very big challenges about the possibility that vibranium is a new element.

[††]On a lab scale in 1989 I achieved an ultimate tensile strength of 729 MPa for a piece of aluminium alloy.

Scientists discover new elements regularly, as they are synthesized by colliding lighter elements together to create heavier fusion products. Unfortunately, these elements tend to have very short half-lives. For example, the most stable isotope (293) of element 116, livermorium, has a half-life of only around 60 milliseconds. To put this in context, if one tonne of this element was created (and let's not worry how to do this ...) then in just 1.2 s less than 1 g of livermorium would remain. There are physicists who propose that there are what are termed 'islands of stability' whereby some higher atomic number elements would have longer half-lives. However, even if these half-lives became significant, the same physicists tell us that these elements would be very dense – very much denser than steel. So, we can be certain that vibranium is not an as yet unsynthesized element.

So, if vibranium is not a high atomic number element ahead of its documented discovery, could it be a well-known element in a new guise? The element carbon provides some encouragement here.

Over the past three decades our understanding of carbon has changed dramatically. I passed my A-level in Chemistry in 1986 by drawing sketches of just two allotropes of carbon, i.e. diamond and graphite. If I had claimed there were more I might have failed. Since then, what I learnt in A-level Chemistry has been overturned by scientists who went on to win Nobel Prizes. One of these celebrated scientific breakthroughs was the discovery of the buckminsterfullerenes. The most famous of these, C_{60}, consists of carbon arranged as pentagons and hexagons to create a nano-football. The lab adjacent to my office regularly makes use of these C_{60} molecules to analyse material. It does this by firing them at the surfaces of materials, as part of a mass spectrometry technique.

Hot on the heels of C_{60} came the carbon nanotubes, which again are a family of possibilities rather than a single material. For example, they come in single-walled and multi-walled forms. It was not long before the realization came that a form of carbon existed which can be viewed as a cut and flattened nanotube, or as a single layer of graphite. Scientists are looking at the properties of what has been named graphene, and have gone on to found a new branch of materials science looking at what are now known as 2D materials. Just a few miles from the Manchester

Science Festival where this book was written lies the University of Manchester's National Graphene Institute founded in 2013, where these discoveries were made.

Could vibranium be graphene? After all, if we look at the graphene infographic (in Figure 7.1), which reports recent nanoindentation data from Lee *et al.*,[2] then we can see that its strength is spectacularly higher than that of the other already strong materials, such as super bainite steel, diamond and boron carbide (see Figure 7.1 also). There has been much debate in the materials science community about just how the properties of graphene can be measured and disseminated. Measuring the properties of a sheet of material that is only one carbon atom in thickness is very challenging. In short this means that we have to be rather cautious about the meaning of its strength because it has to be inferred rather than measured directly. Even if we assume that it has the spectacular strength that has been suggested, there is still a problem with graphene making a super-strong shield.

Graphene is often branded as being 200 times stronger than steel. Yet, as Perkins[3] reports, University of Cambridge metallurgist Sir Harry Bhadeshia has pointed out that such a ratio is based on a number of poor assumptions. A key factor is that two-dimensional materials exhibit something close to their theoretical strength. This concept of strength is based on a material being perfect, that is, free of defects. We have already seen how 3D metals have their strength lowered by the presence of dislocations. A little later we will consider how ceramics have their strength lowered by the presence of microcracks. As we shorten a dimension of a material, statistically speaking defects become less and less likely to appear in our material. If we want to compare our atomically-thin graphene with steel, then we should compare it with iron on a similarly small scale. Whilst it is not possible to make a single layer of iron under normal conditions, scientists such as Brenner[4] researched very thin cylinders of material known as whiskers some decades ago. On this basis graphene is only around six times stronger than steel.

The newly discovered 2D materials, and the even as yet undiscovered ones, cannot explain vibranium. There are very good physical reasons for why their theoretical strength cannot be realized on macroscopic length scales. So, have we reached the end of the road for the scientific plausibility of vibranium? Well, no, there are some further possibilities to explore. We have established that vibranium is not a high atomic number element but that it is metallic, which leaves some further possibilities.

7.6   IS IT A COMPOSITE?

There are ways in which materials scientists can have their cake and eat it when it comes to materials properties. This is possible by combining materials that have complementary properties to create composite materials.

Arguably the most successful composites are a combination of continuous carbon fibres set in a polymer matrix. Such materials are often rather unhelpfully referred to as 'carbon' or just 'composites', but precision is needed as they are clearly more than just carbon and there are other types of composites.

Fibre-reinforced composites are highly tailorable materials. Their strength and stiffness can be altered by changing the fraction of fibres in different directions. The possibilities are huge because thin layers of materials can be combined together such that the final product has various layers which can lie in different orientations. Until recently, composite materials were most likely to be encountered as glass fibre, although they are beginning to be replaced by carbon fibre-reinforced polymer composites. Both of these composite systems combine a high strength, high stiffness but low toughness fibre embedded within a low strength, low stiffness, relatively high toughness matrix. The benefit of this is a class of material which has among the highest specific strength and stiffness of any engineering material, yet at the same time a level of toughness which enables them to be exploited in applications including high-performance racing cars and civil aircraft wings.

Of course, vibranium is not a polymer composite because we have established that it is metallic. So, what of metallic matrix composites (MMCs)? MMCs have made a small impact as advanced engineering materials over the past 20 years or so. Current commercially available composites are mostly aluminium based and are reinforced with particles of a suitable ceramic, which is often silicon carbide.

Reinforcing a metal with a ceramic is an attractive option in terms of the potential for increasing strength. In the infographics (Figure 7.1) we can see that the strengths of diamond, alumina and boron carbide are very high. Although they do not realize strengths close to their theoretical strength they offer potential to improve metals when incorporated as reinforcement. Some ceramic materials exhibit a combination of properties that does make them useful for applications analogous to Captain America's shield. For example, boron carbide is used as armour tiles on tanks. Its compressive strength and hardness make it highly effective at stopping high velocity and high mass projectiles. Where it differs from vibranium is that it is essentially only fit for stopping a single projectile – when it is hit by a projectile it shatters into tiny fragments. When it is used as armour plate on a tank it is tiled, so that the damaged tiles can be replaced.

Perhaps vibranium is a composite material, giving us the best of a metal with the best of a ceramic: tough yet very hard and very strong. Even if this were true, there are still two more hurdles for us to overcome in providing a rational basis for vibranium's property balance. Modern metal matrix composites are strong, but not as strong as vibranium, and we also have to contend with vibranium's amazing energy-absorbing capabilities.

The strongest metal matrix composites that are on the verge of commercialization have a titanium matrix. Titanium alloys such as the alloy in Figure 7.1, which contains additions of 6% aluminium and 4% vanadium by weight, are already high-strength alloys. To improve their strength significantly the ceramic silicon carbide can be incorporated. Rather than using particles, as in commercial aluminium MMCs, continuous fibres known as monofilaments can be used. These monofilaments typically have diameter of a little over 100 micrometres (which is about the same diameter as a human hair). Such a material also has a density that is approximately one-third of the density of steel. A titanium alloy composite made of 35% silicon carbide has a density of around 4000 kg m^{-3} compared to the 2700 kg m^{-3} mentioned earlier. Such monofilaments and alloys are on the cusp of commercialization – see Rix *et al.*[5]

So, despite our hard work thus far, we are struggling to find a genuine match for vibranium – even with some new near-commercial materials. However, the gap is relatively small now that we've met our ceramic-reinforced titanium matrix composite. There is also hope that a fairly new idea in materials science might offer a plausible explanation to bring vibranium firmly into the realms of possibility.

7.7 HIGH-ENTROPY ALLOYS

In 1996, Huang *et al.*[6] examined a key limitation of alloy development. Their work alludes to the fact that, historically, most engineering alloys have been based on a single specific element to which other elements have been added. This trial-and-error approach to alloy development has dominated metallic materials research until very recently. This explains why the most widespread alloy systems are named after the principle element. Thus introductory books on metallurgy will mention aluminium alloys, copper alloys, ferrous (*i.e.* iron) alloys, magnesium alloys, titanium alloys and nickel alloys, leaving the reader thinking there are no other engineering metals. However, there are exceptions to this rather singular focus. When it comes to some alloys of iron, nickel, chromium and cobalt, it is not always obvious what to call them because they contain large amounts of two or three of these elements.

Essentially, Huang *et al.*[6] proposed a new category of alloy in which the incremental tweaking of alloys based on one element is turned on its head. The idea is simple: consider alloys that comprise five elements, all present in large amounts. In a more recent paper, Tsai *et al.*[7] define these so-called high-entropy alloys as 'alloys with five or more principal elements. Each principal element should have a concentration between 5 and 35 at.%'. Such alloys offer enormous potential because they open up an enormous range of possibilities for the exploration of new compositions. Contrary to received metallurgical wisdom, they do not simply lead to the proliferation of metal–metal compounds known as intermetallics, which are very brittle phases and therefore low toughness alloys. Rather the very high entropy of mixing of five or more elements in an alloy results in the formation of solid solutions that can be both strong and tough. As we have already seen, combining toughness and strength is central to accounting for vibranium. The maximization of both strength and toughness is also key to everyday applications such as airframes, aircraft landing gear, biomedical implants, racing cars, production cars, *etc.* If vibranium is a high-entropy alloy reinforced with a ceramic for good measure this offers a possible explanation of its remarkable properties of low density, high strength, high hardness and high toughness.

But what of its vibration-absorbing capacity – which after all gives it its name? Is this the deal breaker for our metallurgical *tour de force?*

7.8   BAD VIBES

The ability of a material to absorb vibrations is a complicated phenomenon, as explained by Chung.[8] The reason that it is complicated is due to the very large number of mechanisms that enable a material to absorb vibrational energy. Different classes of material (polymers, metals, ceramics, *etc.*) differ substantially and this further complicates matters because it has led to the creation of very different methods for measuring the capacity of a material to absorb vibrational energy.

Our concern with vibranium is to appreciate how effective metal matrix composites might be in damping vibrations. The immediate answer is not very encouraging: metals – as a class of materials – have only moderate vibrational damping capabilities. Lead is a notable exception; however, this is because of some specific characteristics of lead which would not be shared by vibranium. One of the routes to damping vibrations is microstructural behaviour. When it comes to lead, this microstructural damping operates by the successive backward-and-forward motion of dislocations. This dislocation motion is slowed by the inherent friction of atoms moving relative to one another (termed lattice friction) and it lessens the vibration with each reversal of dislocation motion. This mechanism is only open to metals with a low melting point which are relatively soft at room temperature. Everything we have done so far is predicated on vibranium being very hard.

There are other more plausible microstructural mechanisms for absorbing vibrational energy. One of these is the presence of interfaces within a material. In the case of our metal matrix composite we have the interfaces between the metal matrix and the particle or monofilament reinforcement. These interfaces will provide a damping effect; however, it will not be at the level necessary to explain vibranium's acutely effective damping capability.

The most effective vibration-damping alloys are a class of material known as shape-memory alloys. They have something close to the extreme capability that vibranium exhibits. Shape-memory alloys damp vibrational energy *via* a reversible crystal structure change. Despite their name, it is not the shape-changing capability that is relevant to vibranium so we will focus on a side-effect of the shape-memory effect known as psuedoelasticity.

As we have seen, metals exhibit toughness and ductility by virtue of their ability to plastically deform. Plastic deformation is normally accomplished by dislocation glide. There are some materials, however, that can change crystal structure when a stress is applied. This is possible when two phases have similar energetic stability (at a specific temperature interval) and their crystal structures are related by a mechanical shear deformation. It is then possible to literally deform one into the other. This plastic deformation can be reversed if the stress is reversed, which gives rise to the term psuedoelasticty.

Figure 7.2 shows a stress–strain response, curve E, for a high-strength psuedoelastic material (perhaps vibranium?). The initial application of stress to this material gives an elastic response. At higher stresses the initial phase, known as austenite, is changed into a new phase. This transformation is gradual with just a tiny plate at a time being transformed. In this way a large amount of energy and stress is required to change the material into the new crystal structure. Eventually the transformation is complete and further deformation will eventually lead to yielding by dislocation glide and the more usual response of a metal now follows. This reversible switching from one crystal structure to another can also provide a very efficient mechanism for the absorption of vibrational energy.

7.9   CONCLUDING REMARKS

So, vibranium could be a high-entropy shape-memory alloy composite, reinforced with a ceramic. The manufacture of a material as sophisticated as this is beyond the limits of current materials science. Moreover, science can offer a rational justification for the performance of Captain America's shield, just about. All of our reasoning makes it clear that such a material is not a natural material.

Native metals do occur, but those that do are near-pure relatively unreactive metals like platinum, gold, silver and copper. How does this square with the Marvel Cinematic Universe's vibranium, which is said to be found only in Wakanda? Well, we are left with the implication that, despite our success in fitting all of the data for vibranium to a high-entropy shape-memory alloy composite, reinforced with a ceramic, vibranium must be of extra-terrestrial origin.

Perhaps in the Marvel Cinematic Universe a spacecraft from another solar system crashed in Wakanda? If a substantial part of its hull was vibranium it would have the appearance of a meteorite of a new supermaterial. I, for one, am looking forward to testing this hypothesis – as would all good scientists. This hypothesis can only be tested whilst ingesting ultra-low density corn sprinkled lightly with sodium chloride in a darkened room with a silver screen.

REFERENCES

1. R. I. Hammond and W. G. Proud, Does the pressure-induced $\alpha \rightarrow \varepsilon$ phase transition occur for low-alloy steels, *Proc. R. Soc. London*, 2004, **A460**, 2959–2974.
2. C. Lee, X. Wei, J. W. Kysar and J. Hone, Measurement of the elastic properties and intrinsic strength of monolayer graphene, *Science*, 2008, **321**, 385–388.
3. J. Perkins, Doubt cast on graphene strength claims, *Mater. World*, 2015, **25**(12), 20.
4. S. S. Brenner, Tensile strength of whiskers, *J. Appl. Phys.*, 1956, **27**, 1484–1491.
5. M. V. Rix, M. A. Baker, M. J. Whiting, R. P. Durman and R. A. Shatwell, An improved silicon carbide monofilament for the reinforcement of metal matrix composites, *TMS2017: Proceedings of Pan American Materials Congress: Materials for Transportation and Lightweighting, 2017 Feb 26–Mar 2*, TMS, San Diego, Pittsburgh, 2017, forthcoming.

6. K. H. Huang and J. W. Yeh, A study on multicomponent alloys systems containing equal-mole elements, M.S. thesis, National Tsing Hua University, Hsinchu, 1996.
7. M.-H. Tsai and J.-W. Yeh, High-entropy alloys: A critical review, *Mater. Res. Lett.*, 2014, 2(3), 107–123.
8. D. D. L. Chung, Materials for vibration damping, *J. Mater. Sci.*, 2001, **36**, 5733–5737.

CHAPTER 8

The Science of Super Suits

SUZE KUNDU

University of Surrey, Department of Chemical and Process
Engineering, Guildford, Surrey, GU2 7XH, UK
E-mail: s.kundu@surrey.ac.uk

8.1 INTRODUCTION

Let's face facts here. The majority of superheroes gained their
superpowers through either chance encounters, or the kinds
of lab accidents that would result in hours of health and safety
paperwork. For me, as a materials scientist and engineer, the real
superheroes that I aspire to become one day are those who are
most like me – regular people with a lab to hand, and a bunch of
money. In my case, that would be hard-earned grant funding. For
them, it is usually an inheritance pay-out or the acquirement of a
large family company. Either way, superheroes like Batman and
Iron Man are hugely inspirational in terms of science and engineering potential. However, the powers that their suits hold are
confined to the realms of science fiction and comic book pages ...
or are they? Time to get my science on and delve into the real-life
materials of super suits, whether we can really recreate them to

The Secret Science of Superheroes
Edited by Mark Lorch and Andy Miah
© The Royal Society of Chemistry, 2017
Published by the Royal Society of Chemistry, www.rsc.org

become superheroes ourselves, and settle the Marvel/DC score of who would win in a battle between Batman and Iron Man. They have already battled Superman and the delightful Captain America, of course.

8.2   BAT FASHION

Let's start by looking at Batman. After a traumatic childhood, during which he witnessed the brutal murder of his parents, Bruce Wayne grows up and vows to fight crime. Not content with sticking a Neighbourhood Watch sticker in his mansion's porch window, he ramps this up several gears and becomes the caped crusader that we know today.

Is the suit just for show, a flashy way to conceal his identity? Maybe a little – as it is a pretty snazzy get up. However, the suit possesses a whole host of amazing properties that help our Brucie both fight those baddies and keep him safe. It is light, robust and strong enough to save Batman from bullets even when fired at close range. In *Batman v Superman: Dawn of Justice*, the suit even has a Superman-repellent coating and the ability to give Batman super strength. What materials could be used to make a suit of such powers?

Many of the iterations of suits in Batman's 'everyday casual' range of elaborately caped onesies are dark grey and black in colour. The different areas of the suits look like they have different textures, and his boots and gloves are carefully designed for so much more than simply meeting the latest Autumn/Winter fashion trends. The density and durability of these suits is also important when considering how cumbersome wearing a suit like this would practically be when saving the world from villains.

Let's start by thinking about the things that can really make or break an outfit – the accessories.

8.2.1   The Bat Gloves

Batman is wearing more than gloves. These armoured beauties fall into the gauntlet category of handwear, and for good reason. Not only do the scallop-finned gauntlets totally adhere to this year's Autumn/Winter fashion trend, they spare Batman from injuries incurred from bullets and swords. The fins can even be fired off and used as projectiles aimed at enemies. Given their propensity to remain undamaged even after such fierce impacts, they must be made of pretty tough stuff, but their amazing abilities don't end there. They are shockproof and resistant to radiation. Meeting all of these challenging properties in one material can be tricky, but this is made easier by using a combination of materials with these desired properties and designing a material in such a way that said desired properties of materials are not compromised by one another's presence. Given that the gloves also need to be light enough and flexible enough for Batman to not compromise his best fight choreography, these materials need to be strong but have a low density. However, when we talk about a material's strength, we can mean a range of things, and so it is important to explore this term first.[†]

The strength of a material is its capacity to withstand different forces and return to its original shape or size without breaking or getting squashed. This includes a material's capacity to not dent or become otherwise deformed in shape, not break, not get scratched, and also withstand extremes in temperature and pressure, and not fall apart after just a handful of uses. There are various ways of measuring a material's tensile strength, from pulling it apart, to measuring a material's modulus of rupture using a three-point-bend test to see how much force it can withstand before fracturing or bending out of shape. It's also possible to measure its robustness by poking it with an indenter.

In the case of Batman's gauntlets, they must be made of a material with high tensile strength, so that it does not tear with the impact of a weapon, with the addition of being flexible enough to wear, and elastic enough to absorb and dissipate the energy of the impact to save him and his wrists from the shock, while also saving him from radiation. A material like graphene,[1] the wonder material that is said to fix almost every problem, could certainly meet the challenge of strength with its huge tensile strength of 130 000 MPa (mega pascals – the unit of measurement when discussing strength in this way). Graphene is also flexible, allowing Batman to fight unimpeded.

[†]This concept crops up a few times throughout the book. We look at it in much more detail in Chapter 7 on vibranium.

The problem is that graphene only exists in a two-dimensional single layer of carbon atoms, and this is not only difficult to work with, it is also very difficult to manufacture in a large enough area for it to be used in this way, and we are still not 100% sure about how it may interact with the human body, owing to its nano-dimensional structure.

While carbon nanotubes are also a contender in this category, they are also subject to the same concerns, despite their high tensile strength of up to around 60 000 MPa. However, nanotubes lend themselves more easily to their use in composite materials. What else could work in this instance? More traditional, lower tech, solutions like leather are not bad,[2] with tensile strengths of up to 27 MPa. However, there are stronger synthetic materials available on the market that are safe for human use, and which can be further strengthened by weaving other materials – like graphene – into them in such a way that the resultant fabric is also very tear resistant. Kevlar® is a great example.[3] With a tensile strength of 3620 MPa and an increased resistance to tearing,[4] the additional favourably low density of Kevlar® makes it an ideal candidate for use in these ultimate safety gloves.

What about addressing the need for the gloves to also be shockproof and radiation resistant?

Kevlar® is also great at dissipating the energy of an impact quickly across a wider area – one of the reasons that it is so extensively used in bulletproof vests – but to increase this property, it could be possible to incorporate an additional material with these energy dissipation skills like non-Newtonian D3O. Unlike a regular 'Newtonian' material that behaves like a normal liquid with regular viscosity regardless of the forces acting upon it, the particles in this bright orange slime can lock together on impact, briefly solidifying and more efficiently dissipating impact energies, but after impact return to being a comfortable, light, flexible material. That may sound like a very high-tech super-material, but you can see exactly the same thing happen when making gravy. A thick suspension of cornflour also behaves like a non-Newtonian liquid (find yourself an 'oobleck' recipe and have a go). Cornflour is a bit tricky to incorporate into a material, so instead D3O is often embedded into foams which can be used to line regular items of clothing to provide a level of impact protection, such as non-Newtonian beanie hats for use as crash helmets for snowboarders.

As for radiation protection, it really depends on the type of radiation that is being blocked out. Generally speaking though, lining these gloves with a layer of lead may be enough to meet this requirement, but more on this later.

8.2.2 The Bat Boots

Back to Batman's arsenal of accessories, and specifically a group of accessories that I know a fair bit about; boots. Not only are the Bat boots incredibly chic, they are made of materials perfect for any terrain. It is likely that these boots are made from a reinforced rubber. Natural rubber mostly contains a molecule called *cis*-isoprene, which contains loads of carbons and hydrogens, and therefore falls into the hydrocarbon group of organic molecules. Rubber is elastic and waterproof, but its grippy properties arise not only from the molecule's propensity to form weak attractions between the rubber and a surface, but also the physical composition of the material. It is both porous and flexible, allowing its uneven grooves on the surface to sink into the dents and ridges of an uneven surface with a huge surface area. It also has a high coefficient of friction, a material property which tells us how much friction would be produced between two specific materials. Combining both the high surface area and the high coefficient of friction of rubber, we get a very large amount of friction, and therefore a high degree of grippiness.

To increase this further and make it adaptable to a range of soft and hard surfaces, grooves are cut into the soles of the boots. The boots are 'split sole'; there is an area of grip on the ball of the foot and the heel of the foot, but an absence in the middle where the arch of the foot is. Not only can this offer superior arch support and a higher degree of flexibility when perched precariously on a ledge, the jazz dancer's staple of a split sole shoe would allow Batman to point his toes like Darcey Bussell, should he ever choose to partake in a musical theatre number while fighting crime. To combat wear and tear, the rubber boots are likely to be reinforced with our old friends the carbon nanotubes, which we briefly encountered earlier when discussing possible glove material. Their tensile strength and random orientation within the reinforced rubber composite give the material greater robust-ness and durability on rough terrains, and also ensure they don't wear down mid-combat – this is Batman, not Die Hard. Bare feet are highly unnecessary. While the rubber composite probably used to make these is flexible and comfortable to wear, they offer little protection for Batman's feet should he choose to engage in foot-first fighting. To pack more of a punch – or a kick, in this case – the toes of his boots are highly likely to be steel-capped inside. You would not want to be on the receiving end of one of these boots!

8.2.3   The Batsuit

Accessories out of the way, let's delve into the science of the basic Batsuit. This bodycon body armour needs to be light and tough. There would be no time for any Beyoncé-style costume changes for Batman. His outfits have to last the long haul. In the comics, Batman's suits are bulletproof and tough to tear. They are also able to insulate him from electricity, and also dampen the effects of shocks and impacts. While the shock-absorbance properties could be provided by a layer of D3O foam within the suit and the bulletproof and tear-resistant properties by a Kevlar® weave, we need to add in another material to address the electrical issues. This material needs to be electrically conductive, so that an unin-terrupted layer or continual mesh of this material can conduct electricity through the Batsuit itself, and around the contents of the Batsuit, protecting Batman inside. This concept is known as a Faraday cage, named after one of my favourite scientists Michael Faraday. He was one of the scientists to develop this idea of being able to protect something from the harmful effects of electromagnetism within a cage.

We see Faraday cages being used in microwave ovens. Look closely at the see-through door of a microwave oven and you will see a mesh or cage with small holes within it – this stops the microwaves from escaping. We even use this idea in aeroplanes, which are made of materials that can conduct the electrical charge of a lightning strike around the body of the plane, protecting the passengers within the aeroplane. Good conductors of electricity include pure silver, pure copper and pure gold, but these mate-rials are incredibly expensive (so much so in copper's case that electrical wiring is frequently stolen by people trying to sell it on at a high price[‡]). They are also good at conducting something else; heat. If you have ever accidentally touched an incandescent light-bulb that is on or has recently been turned off, you will know that heat is generated when electricity conducts through a material. In a lightbulb's case, heat is given off when electricity flows through a small and very thin wire made of a material such as tungsten. So much heat is generated that the wire glows brightly enough to light a room, which is how lightbulbs work.

If Batman's suit used gold to create the protective Faraday cage, it could also heat him to uncomfortably hot temperatures – and no one needs that when they are fighting crime in a tailored onesie. Instead, we can look beyond the classic metal conductors and back to our old favourite, carbon. Yes, the periodic table's resident overachiever delivers once more, this time offering up two potential solutions. We have already encountered graphene in terms of its strength and lightness, but a graphene cloak embedded within the Batsuit could conduct any electrical shock around Batman, sparing him from the nasty effects of electro-cution. Graphene is an excellent conductor of electricity, and although it also a very good conductor of heat, it only does so along the direction of its planar two-dimensional sheet. It is also incredibly light, given that it exists in an atom-thick layer, which would prevent this additional bit of technology from weighing Batman down. In reality graphene can be very difficult to pro-duce in the sorts of large continuous areas that would be vital for the suit to protect Batman from electric shocks. However, nano-dimensional carbon could be the solution after all – and the best news is that it has already been road-tested by a physi-cist in California!

Dr Austin Richards works in the area of infrared imaging sys-tems, and, back in 1981, initially started off making Tesla coils, which can transfer huge amounts of electricity wirelessly through the air in beautiful arcs of 'lightning' for fun. Subsequently, he moved on to making his own Faraday suits in 1997, allow-ing these huge electrical charges to pass around him while he

[‡]An incident which once left me stranded at Reading Rail Station on my way back from Glastonbury Festival as someone had stolen the copper overhead power cabling ... Thank goodness for the bookshelves in WH Smith!

effectively mimics another superhero and Norse god, Thor, and plays about with lightning. Dr Richards has showcased his suit in many places where he plays the part of Dr MegaVolt. Previous gigs have included Burning Man Festival in California (a bit like the aforementioned Glastonbury, but with more sunshine, less rain, and no unplanned visits to Reading Rail Station), where he rigged two trucks to fire arcs of electricity at him, and lived to tell the tale.[5] So, how does his Faraday suit work?

Graphene may be tricky to work with, but roll some sheets of it up and you have a new material that can also conduct electricity quickly and vectorally along the length of these straws, called carbon nanotubes. These were mentioned before, but not in much detail. Essentially they maintain many of the desirable properties of graphene, but exist in a more easily obtained way. They are synthesized naturally in the soot given off when wood is burned, but can also be synthesized in a lab. If these nanotubes are woven together, a strong material known as carbon fibre is produced. If a suit is embedded with this light and conductive material, the wearer can be protected from all electric shocks as the carbon fibres all overlap within the suit, conducting the electricity around the wearer's body, and saving them from a shock. Clothing like this can stop anyone from getting electrocuted or tasered, but only if the electrical charge is applied to the clothing itself. If exposed to skin, the wearer will feel it. Luckily for Batman, he has his neck, hands and face well covered. With carbon fibres being incorporated with Kevlar® fibres to make Batman's suit, he is protected from shocks and shots, without being weighed down by his wearable tech.

Though Batman is well protected from electric shocks, he does have the power to administer them as a defence mechanism, thanks to his electrifying cowl, or hood. It is possible that he has incorporated some form of tiny Tesla coil within his hood, though when these were invented in 1891 by Nikola Tesla they were supposed to simply transfer electricity quickly without the need for wires. Tesla coils work with high-voltage electricity, which we do not normally encounter. Tesla coils are so-called because they are made of two coils (primary and secondary) and two capacitors (also primary and secondary, one for each coil). The capacitors are capable of storing lots of energy until they are ready to release it all. By taking low-voltage mains power, and ramping up its voltage with the use of a transformer, the primary coil is linked to the power source, and the primary capacitor guzzles up as much electrical charge as it can handle, before quickly discharging this through the primary coil by bridging a gap in the circuit known as a spark gap, which can only happen if enough charge has built up. As this electricity quickly flows through the primary coil, a magnetic field is generated, which electrically speaking gets things moving in the secondary coil. This electrical charge starts to then build in the second capacitor, until it can hold it no longer and it cracks through the air like a bolt of lightning. While this would not lose any energy in a perfect system, heat is lost through the arc of lightning, and so the Batsuit hood always needs to be plugged into a power source in order for this to work on demand, limiting its uses somewhat. Could high-powered batteries work instead? Well, if we take Dr MegaVolt's party trick as an example, he uses a battery pack to generate arcs of 4 inches. These battery packs can last a while without being recharged, so providing Batman is not going into prolonged battle, these could be the answer to making his electrifying hood.

8.2.4 The Bat Hood

Batman's hood is multifunctional, allowing him to spot danger, thanks to its scanning capabilities, ask Alfred to order him a take-away, thanks to the ear transmitter, and stay safe and mysterious, thanks to the lead-lined Kevlar® panels that prevent him from getting injured, and stop anyone from X-raying his face to reveal his true identity. The first two of these functions require energy, so his suit may need to be capable of harnessing or generating energy in one form or another. He could make the most of the large area of his suit or cape and embed solar harvesting materials into it, powering his battery pack up as he goes, so that he can keep in touch with everyone and still have enough power left to deliver a shock to anyone that gets on the wrong side of him.

8.2.5 The Bat Cape

So far, I have discussed some of the technology and materials engineering that could explain how Batman's suit is possible, but we have neglected to mention the pièce de résistance; the cape, said to be made from a material called 'memory cloth'. If you thought his hood was a multitasking marvel, allow the cape to blow your mind. It is fireproof, bulletproof, light, tough, allows him to glide and has a high potential swishability – all good qualities of a cape, I think you'll agree.

Of all the features of the Batsuit, some materials have cropped up numerous times as potential candidates to make this suit a reality, often due to a different one of their myriad properties. However, there is a great analogy for science itself here – things are better when they work together. Collaboration is key to success in science and engineering, and materials science is no different. By combining two or more materials appropriately, a material can be created that has a wider range of properties and an enhanced ability to meet those requirements. Kevlar® is tough to tear and difficult to burn, but is susceptible to folding, which is the last thing you need as you are gliding through the sky from building to building. In order to maximize drag in the air, the cape needs to hold its shape and spread out as broadly as possible. By combining two materials and weaving them together to form a flexible fabric, we could have our low density, tough, fireproof and bulletproof cape material that can, on command, morph into a shape optimal for gliding. The crucial material in this mix is a form of shape-memory material.

Shape-memory materials are materials that are set into their desired shape, usually using high heat or chemical treatment, but at room temperature can be manipulated into any other shape. When the desired shape is required, the material can be activated by a change in temperature, the presence of light, or most likely in the case of the cape, through a short burst of an electrical current. Materials like these are already in use in everything from arterial stents to the underwiring of bras. So-called electro-active shape-memory materials[7] would be ideal for Batman's cape, especially given that he is already carting around some significant circuitry. They can be made of metal alloys such as Nitinol (nickel and titanium alloy) or a delightful blend of other materials, such as carbon nanotubes and a graphene aerogel[6] – honestly, for a self-proclaimed inorganic chemist, I can't seem to stop figuratively fangirling over carbon-based materials. They can also be made of polymers, which would be preferred in this case as they are much lighter than the same volume of metal alloy. One problem with shape-memory materials is that they can lose their ability to spring back into shape as effectively over time; however, a team of scientists discovered by serendipitous chance a material that almost entirely resists degradation when repeatedly bent and sprung back into shape[8] – handy, given that these materials are still pretty expensive, and you wouldn't want to have to keep sewing that cape up, especially with Alfred getting on a bit.

8.2.6   The Bat Supersuit

So far, all of this science has gone into making a pretty standard Batsuit, but what about the monstrous exoframe we encounter in *Batman v Superman: Dawn of Justice*? In this suit, extra steps are taken to weaken Superman while also boosting Batman's strength. We all know that Superman's Achilles heel is the mineral Kryptonite. Batman uses this against Superman in Dawn of Justice's battle in the form of releasing Kryptonite grenades and a Kryptonite-tipped spear, but he also manages to pack a much stronger punch than his size and mass would be capable of delivering. How does he manage it?

The answer is found in artificial muscle – materials that can contract and relax on demand, just like a real muscle, only these are embedded within the suit itself, can contract more rapidly than human muscle, and are much stronger too, requiring only a small input of energy. Current research into artificial muscle is focusing on a range of materials, such as polymers and shape-memory alloys. One particular material[9] is very interesting, as it is over 200 times stronger than human muscle of the same dimensions. Made of paraffin-filled carbon nanotubes that have been twisted together to form carbon fibres, this nano-maccheroncelli rapidly contracts when heated as the paraffin expands, and the carbon nanotubes stretch out sideways to accommodate the extra volume taken up by the wax inside them, pulling the ends closer together in the process. When the heat is removed, the paraffin shrinks again and the nanotubes return to their relaxed and elongated shape once more. If Batman's epic exoframe is embedded with a material like this, then a tiny increase in temperature could switch them on, and given the fact that they are 200 times stronger than human muscle, it is no wonder that he can take on an alien with super strength. Artificial muscles have been a subject of much research interest for a few decades now, but scientists are starting to put a lot of this research into practice, giving hope for people reliant on prosthetics. If these can be enhanced with artificial muscle, it could provide the wearer with increased mobility. Furthermore, scientists are even looking into materials that can behave like heart muscle, with the ability to pump a liquid around a system.[10]

8.3   IRON MAN

The antics of a certain Tony Stark are pretty far-fetched. With more family money than any research council could ever bestow upon a scientist or engineer and without having to worry about any dull university bureaucracy, Stark has the freedom to create and the bank balance to experiment. An established businessman, inventor and entrepreneur, Stark is captured and suffers an injury when a piece of shrapnel enters his body and moves dangerously close to his heart. Under instructions to build weapons of mass destruction by his captors, Stark instead builds a suit to escape with, and a device that keeps the metal away from his heart. He does have heavy contributions from a fellow captive that luckily happens to be a Nobel Prize-winning physicist – who sadly dies as Stark escapes.

8.3.1   Not-Iron Man

And so began the range of Iron Man suits[11] created by Stark, many following the standard 'protect, survey and defend' model with the ever-present arc reactor saving him from his shrapnel injury and powering the suit, and some also with some very interesting capabilities created for very specific applications – hello, Hulkbuster – but all rather incorrectly named.

Although the original suit of armour was cobbled together from iron-containing steel, future iterations must be made of other materials to allow them to work. Not only is iron incredibly dense, it is also prone to denting, bending and rusting. One material that could stand up to the torrent of abuse received by one of Iron Man's suits is titanium. It is much lighter than iron or even steel, is tough and durable (though not as strong as the toughest steel), but can withstand both extremes in temperature. It is also incredibly difficult to get it to react with other chemicals like acids, alkalis and water. So far, titanium is a better option than steel or iron if we are talking about rigid suits of armour, but it may not be able to deliver all of the properties that Iron Man's many suits possess.

First of all, Iron Man's modern suits do not just exist as sheets of metal. They are assembled from scratch in a matter of moments from smaller component parts about the size of salt grains. Self-assembly of materials does exist in nature – the mere fact that we are even here writing, or in your case reading, this chapter is evidence of that. But materials scientists are on a quest to take inspiration from nature and find artificial materials that can do the same. In doing so, much like the Iron Man suit, we can create working machines from component parts on a tiny scale. Imagine being able to inject the building blocks of something the size of Ant Man into your body, and stimulate it with the correct conditions to trigger assembly inside you so that some routine surgery or diagnosis can be carried out with minimal invasiveness.

The components of the early versions of the suit could assemble themselves in seconds. Each of these so called 'cells' is basically a tiny machine in itself, linked to every other one by a high-powered force field, rather like batteries connected in a circuit. However, this 'circuit' is epic, as every cell can gather its own energy and work independently of the rest of them, while still being in contact with all the other cells. This reminds me a little of the quorum sensing that bacteria perform, pooling their individual powers to act as one more powerful entity that takes action to achieve a shared goal. This sort of material could potentially give rise to more shape-shifting properties of the suit, possibly allowing Tony to turn into a Transformer!

Many of the suits are great at withstanding the impact of physical attacks. They are also able to launch missiles, which could have quite an impact on the suit when we consider Newton's third laws of motion. As every action has an equal and opposite reaction, the force of a missile would also rebound on the suit, so the material it is made of cannot afford to be brittle or weak. Given that Iron Man manages to stay tethered to the ground, rather than being thrown backwards, every time a missile is launched, we must assume that the soles of his shoes are coated with a rubbery material with a very high coefficient of friction, as otherwise he would be getting thrown all over the place – something that only tends to happen when he crosses anyone, romantically or otherwise. The suits do not dent as such, and if they do they have the ability to bounce back into their set shape, much like shape-memory alloys are able to do. Perhaps many of his suits are made of sheets of Nitinol, the same material that may be found in Batman's cape. The suits are also bulletproof, suggesting that they are made of very tough materials. It is possible that there may be a reinforcing material present in the suit material composite. When you think about some of the artificial objects that are capable of withstanding a beating in nature, such as an aeroplane in a hailstorm, these are also made of light but strong materials. Traditionally, aeroplanes are made of an aluminium alloy but now also use composites that contain carbon fibre-reinforced plastic.

8.3.2   Insulating a Superhero

A key feature of Tony Stark's more modern suits are the jet boots and the palm repulsors. These emit high-powered fiery jets that can be used to help Iron Man fly, but are also used in defence against enemies, particularly the palm repulsors. These look like little rocket boosters, and must be as powerful to allow him to escape Earth's gravity and actually fly.

Rocket engines reach temperatures as high as 1200 °C, so the suit needs to have some great insulation to keep that heat away from Tony's skin and the delicate electronic components of the suit. Perhaps the suit also needs some great thermal conductors to whip that heat away from those components before they cause any damage. This two-pronged defence mechanism could be achieved by packing certain areas of the suit with a material like an aerogel.

Aerogels are solid foam-like materials with many very fine holes in the material. Aerogels are by definition highly porous, consisting of over 98% air by volume, making them incredibly light. Air is a mixture of gases, so any molecules in air are spread out over great distances. The molecules are also travelling relatively quickly compared to liquids and certainly solids, and so it is very difficult for them to pass on heat through conduction, as they simply do not bump into each other to transfer the heat energy often enough. This makes gases in general good insulators, and aerogel provides a means to apply these insulating properties as a solid material.

Although aerogel is a solid, it can be made more versatile by incorporating small pieces of aerogel within fibres or fabrics. In doing so a new material is created called a pyrogel, maintaining many of the insulating properties of aerogel but in a more user friendly format. Perhaps certain areas of the jet booster boots and the palm repulsors are insulated with such a material as a form of heat protection. Meanwhile, to keep the electronics well protected, it is entirely possible that the suit also contains highly ordered graphite, which maintains some of the excellent thermal conduction of graphene, in a more practical material that has been arranged in such a way that heat is whipped away from the hot boosters and guided away from the precious electrics.

In terms of powering the suit, we all know about the arc reactor that was initially designed to keep shrapnel from piercing his heart. This is also energetically powerful enough to keep his suit fully functional, but some versions of the suit also have the ability to harness solar energy, to keep him going in between recharges. It would be relatively easy to achieve this, though the efficiency would be low – either by applying a layer of a functional material to the outside of the suit, or if the bacterial assembly is to be believed then perhaps by somehow genetically modifying the bacteria-like cells that form the suit to be capable of capturing sunlight and storing it.

8.3.3 How to Pick a Suit

While the more basic suits have been pulled apart and 'scienced' so far, it is important to point out some of the quirkier properties of certain suits. The 'Stealth Armour' suit is black for camouflage, and contains in-built coolers to prevent detection by heat scanners. The 'Space Armour' suit was made of far more durable materials, with in-built life support mechanisms and a better propulsion system. There was even the vanity suit that revealed, or should I say 'displayed', his chiselled features, which we discovered had a faceplate made of tough and transparent graphene[12] – a fact only revealed when Tony was shot in the face at point blank range. Having discussed Batman's hefty exoframe though, it is only right that we bring to the table Iron Man's Hulkbuster range.

A Hulkbuster is basically an exoframe in itself, with the force and durability to stand up to the Hulk when he gets all 'SMASH' in everyone's face. It slots over an existing suit, adding strength (enough to lift 175 tonnes, when an everyday suit could lift no more than 10 tonnes – imagine the shopping possibilities ...) and power to protect Iron Man from Hulk's punches while also dishing some powerful hits back, but all at the cost of manoeuvrability. It is only supposed to be deployed when the Hulk needs to be restrained or controlled, but it does offer a great deal of potential for defence against other enemies, providing they aren't too quick and nimble, as they could easily outwit a hefty Hulkbuster.

8.4   THE REALITY OF SUPERSUITS

Given that Batman and Iron Man both have everyday suits and enhanced suits, if the two were to come up against one another, I can't help but personally say that based on the evidence and my fairly unbiased opinion I would have to say that Iron Man's suit is best equipped for the everyday dangerous situation, but Batman's exoframe wins it for me in the enhanced states. (Full disclosure: I have pledged my allegiance to Marvel for my love of my fiancé Karl, who you will meet in Chapter 14 of this book, and who is sat next to me as I type this, complaining about my Captain America screensaver. Well, if you will make me choose a brand ...) But does it really matter? Surely this is all confined to the realms of comic books, movie franchises and science fiction. Well, actually, no.

A range of private companies, some countries' defence agencies, and some forward-thinking and collaborative scientists have come up with a few examples of supersuits that actually exist.

- The Raytheon XOS 2 Exoskeleton[13] has been developed for the US Army, and increased the endurance and strength of the soldier wearing it, while also being fairly light, very robust, and flexible enough for the soldier to react to any situation.

- Meanwhile, the US Army have also developed TALOS[14] the Tactical Assault Light Operator Suit, that has harnessed some of the properties of non-Newtonian materials to make it comfortable and flexible to wear, but also bulletproof. It is packed with other nanomaterials, and even comes with its own Jarvis-esque health-checking on-board computer, and could be available as early as 2018.
- The Robotics and Human Engineering Laboratory at Berkeley in the USA are also working on an exoskeleton called HULC®,[15] which stands for Human Universal Load Carrier, which enables the wearer to exert less energy when performing what would be otherwise exhausting tasks, such as hiking for long distances or walking up a mountain, while also allowing the wearer to carry a larger load.
- The Warrior Web[16] being developed by DARPA (Defense Advanced Research Projects Agency) is an underlayer that soldiers can wear to ease musculoskeletal strain on their joints when running, crouching or carrying heavy goods.
- Ekso Bionics[17] are a company specialising in wearable robotics that can help paralysed people stand and walk, while also being used to enhance the strength and endurance of soldiers in the field, and were actually the original creators of the aforementioned HULC®.

These suits have been in development for a couple of decades at least, mostly driven by the military's need to protect their soldiers from harm, and allow them to weather tough conditions for longer, but they have undoubtedly been inspired by the imagination and scientific knowledge of comic book writers for several decades. The fact that these suits are now becoming a reality is great, but the fact that they could soon improve the lives of so many people is what makes superhero science so exciting. There is of course still a long way to go to optimise these supersuits and find appropriate materials to overcome the challenges of durability and cost, but that is the beauty of scientific research. Curiosity drives us to strive to find new things out there, and find exciting ways of using them. And who knows – perhaps many of the other superpowers included in this book will mean that, as David Bowie famously sang, we can be heroes, just for one day.

REFERENCES

1. C. Lee, X. Wei, J. Kysar and J. Hone, Measurement of the elastic properties and intrinsic strength of monolayer graphene, *Science*, 2008, **321**(5887), 385–388.
2. C. Ockerman Hansen, *Animal By-Product Processing & Utilization*, Technomic Pub. Co., Lancaster, PA, 1st edn, 2000.
3. J. Quintanilla, Microstructure and properties of random heterogeneous materials: A review of theoretical results, *Polym. Eng. Sci.*, 1999, **39**(3), 559–585.
4. http://www.dupont.com/content/dam/dupont/products-and-services/fabrics-fibers-and-nonwovens/fibers/documents/DPT_Kevlar_Technical_Guide_Revised.pdf, 2016.
5. *History–Dr MegaVolt*, 2016 [cited 1 November 2016], available from: http://drmegavolt.com/history/.
6. *Tesla Coils–Dr MegaVolt*, 2016 [cited 1 November 2016], available from: http://drmegavolt.com/coils/.
7. X. Liu, H. Li, Q. Zeng, Y. Zhang, H. Kang and H. Duan, *et al.*, Electro-active shape memory composites enhanced by flexible carbon nanotube/graphene aerogels, *J. Mater. Chem. A*, 2015, **3**(21), 11641–11649.
8. C. Chluba, W. Ge, R. Lima de Miranda, J. Strobel, L. Kienle and E. Quandt, *et al.*, Ultralow-fatigue shape memory alloy films, *Science*, 2015, **348**(6238), 1004–1007.
9. M. Lima, N. Li, M. Jung de Andrade, S. Fang, J. Oh and G. Spinks, *et al.*, Electrically, chemically, and photonically powered torsional and tensile actuation of hybrid carbon nanotube yarn muscles, *Science*, 2012, **338**(6109), 928–932.
10. S. Kundu, *Forbes Welcome*, Forbes.com, 2016 [cited 1 November 2016], available from: http://www.forbes.com/sites/sujatakundu/2016/03/30/artificial-heart-muscles-one-step-closer-with-cheap-robust-silicone-elastomer-and-carbon-nanotubes/#688e717f2767.
11. S. Perry, *Iron Man Armor: A Complete Guide at SuperHeroHype*, SuperHeroHype, 2016 [cited 1 November 2016], available from: http://www.superherohype.com/features/371529-iron-man-armor-guide#/slide/1.
12. J. Kakalios, *Why Iron Man Ditched Iron for Graphene*, WIRED, UK, 2016 [cited 1 November 2016], available from: http://www.wired.co.uk/article/bulletproof-material-graphene.

13. *Raytheon XOS 2 Exoskeleton, Second-Generation Robotics Suit*, [Internet], Army Technology, 2016 [cited 1 November 2016], available from: http://www.army-technology.com/projects/raytheon-xos-2-exoskeleton-us/.

14. V. Lanaria, *U.S.Military To Deliver Its First Bulletproof, Weaponized Iron Man Suit In 2018*, Tech Times, 2016 [cited 1 November 2016], available from: http://www.techtimes.com/articles/92478/20151007/u-s-military-to-deliver-its-first-bulletproof-weaponized-iron-man-suit-in-2018.htm.

15. HULC™, *Berkeley Robotics & Human Engineering Laboratory*, Bleex.me.berkeley.edu, 2016 [cited 1 November 2016], available from: http://bleex.me.berkeley.edu/research/exoskeleton/hulc/.

16. C. Orlowski, *Warrior Web*, Darpa.mil, 2016 [cited 1 February 2017], available from: http://www.darpa.mil/program/warrior-web.

17. Ekso Bionics, Eksobionics.com, 2016 [cited 1 November 2016], available from: http://eksobionics.com.

CHAPTER 9

Why Doesn't the Invisible Woman Bump Into Things?

KAT DAY

the chronicle flask, chronicleflask.com
E-mail: kat@chronicleflask.com

Invisibility is perhaps the most intriguing and potentially useful superpower of all, but being invisible whilst still retaining the power of sight is actually rather tricky. How does Susan Storm, the Invisible Woman, do what she does? Could refraction hold the key? Is she perhaps bending light waves with some sort of gravitational field? Might adaptive camouflage be the answer? Or is it simple psychology? In this chapter we will take a look (pun intended) at the fascinating science of invisibility, and answer the question: why doesn't the Invisible Woman keep bumping into things?

Ask any group of people what superpower they'd choose, and it won't be long before someone says invisibility. Flying is cool and everything, but someone would eventually see you, and it would only be a matter of time before you ran into difficulties with air traffic control. Super strength is also pretty good, but once you have lifted a few cars, what more can you do with it? X-ray vision

The Secret Science of Superheroes
Edited by Mark Lorch and Andy Miah
© The Royal Society of Chemistry, 2017
Published by the Royal Society of Chemistry, www.rsc.org

sounds all very nice in theory, but do you really want to see the majority of the population's skeletons?

But invisibility, now that has actual, real uses. Forget the moral questions for a moment and just consider the possibilities. You could walk into any shop and help yourself to whatever you fancy. You could even walk into a bank and help yourself by dodging your way into the vault and walking out with the loot. You could watch movies being filmed, get into concerts for free, sit in on lectures from the world's brightest and best, sneak onto an aeroplane and go anywhere you like – although, you might have to stand up. It probably goes without saying that anyone with voyeuristic tendencies could easily get an eyeful of their preferred gender with no danger of being caught. You could enjoy a lucrative career as a hired assassin, among many other things. Of course, each of these possibilities does cause us to consider novel ethical conundrums. We place a lot of importance on the idea of the witness, and it may compromise fundamental human rights if you use your spying capacities on people without their consent.

Even if you have a hankering to be one of the good guys, there are plenty of options. You could pop up behind someone who has taken a hostage and disarm them. You could grab a person who is about to jump off a bridge and save their life. The possibilities are endless – but it is important to remember that being invisible does not necessarily come with the additional power of being able to move through objects. Also, we would have to be pretty clear on whether or not objects become invisible when you pick them up. If not, then some of these abilities might become slightly more conspicuous.

Yet, the real question we have to consider is whether invisibility is actually possible. To discover this, let's consider the most famous invisible superhero of them all, Susan Storm, aka the Invisible Woman. According to Marvel mythology,[1] she obtained her powers after being exposed to a cosmic storm – the traditional radiation exposure, post-atomic age type of thing. Her invisibility is somehow conferred by bending light wavelengths with the power of her mind. Furthermore, she does this without causing any visible distortion effects. And, because the writers did think about the physics a little bit, she also directs enough undistorted light to her eyes to retain her full range of vision while invisible.

This is important because, ironically, vision might just be the biggest stumbling block when it comes to invisibility. Here's the problem: in order for us (or anything else) to see, light has to enter our eyes and hit our retinas. Our retinas contain a mass of light-sensitive cells called photoreceptors, which effectively change light signals into electrical signals. If something is invisible it implies that light is passing through and not interacting with it. If this were the case, those photoreceptive cells would not be absorbing light and we would not see at all.

Science fiction writer H. G. Wells understood this, and wrote things into his 1897 novella *The Invisible Man*[2] to account for it. In his book, the Invisible Man – also referred to as Griffin – makes himself invisible by lowering his refractive index to that of air. As Griffin himself explains in the book:

> Visibility depends on the action of the visible bodies on light. Either a body absorbs light, or it reflects or refracts it, or it does all of these things. If it neither reflects nor refracts nor absorbs light, it cannot of itself be visible. [...] A transparent thing becomes invisible if it is put in any medium of almost the same refractive index.

Refractive index is a measure of how much light bends when it passes from one medium to another. Figure 9.1 shows this for light travelling through air, then through a glass block and then back into air. We can see ordinary glass, even though it's transparent, because

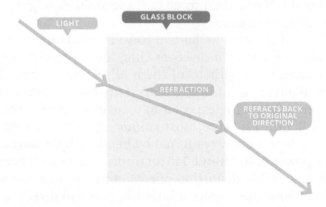

Figure 9.1   We see transparent objects because they bend light. © Andy Brunning 2017.

light bends as it passes through it. This changes the angles at which the light hits our eyes, and since our brain knows that light usually travels in straight lines, it tells us that there is something there. Our brains process this as seeing the glass, which is a good thing, otherwise unloading the dishwasher would be very difficult indeed.

However, if you put a transparent object into something which bends the light in a similar way as it does, in other words into something with the same refractive index, there's no change in the direction of the light for our brains to detect. The object appears invisible. This trick is easy to try for yourself: all you need to do is put a glass rod into a larger glass container which has been filled with clear vegetable oil, and the rod will immediately disappear.[3] Figure 9.2 gives examples of refractive indices for a range of substances.

In *The Invisible Man*, Griffin explains that humans are mostly made up of transparent, colourless tissue. Griffin more than most, since he's an albino. His biggest problem is the red colour in his blood, but he finds a way to chemically remove that colour without changing the blood's ability to transport oxygen. He then manages to lower his refractive index and, after suffering through some unpleasant side-effects, he becomes invisible. However, there is one element that prevents his being completely invisible. When he looks into a mirror, he sees some pigment

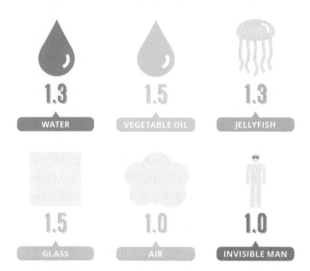

Figure 9.2 Some example of refractive indices. © Andy Brunning 2017.

behind the retina of his eyes, 'fainter than mist'. A similar thing happens earlier in the book, when he experiments on a cat. Its tapetum – the reflective area at the back of a cat's eyes – remains unaffected by the whole process so 'there remained two little ghosts of her eyes'.

H. G. Wells knew that, without this detail, there's no way that Griffin, or the cat, would be able to see. The retina *must* be able to absorb light and something which is absorbing light cannot also be invisible. The Invisible Man would have had bigger problems than this when it came to vision, of course. If the lens in front of his eyes had the same refractive index as air, he wouldn't be able to focus – everything would be blurry. His eyes would also have been overwhelmed by light flooding in from all directions. Wells may have thought of this though: his character often wears thick, dark spectacles (sometimes described as goggles) and prefers to spend his time in darkened rooms.

Back to Susan Storm, the Invisible Woman. She isn't using a refractive index trick to make herself invisible. Instead, she's using the power of her mind to bend light around herself so that it doesn't touch her in the first place, except for the little bit that's directed into her eyes.

But, doesn't that leave us in the same place as in the Invisible Man? If she's directing some light into her eyes, then wouldn't an observer be able to see where that light is refracted and reflected by her lenses, corneas, sclerae (the white parts), irises, and her dark retina behind her pupils? In short, wouldn't we be able to see two disembodied eyes?

This doesn't seem to be depicted in the comics or the movies. Perhaps those scenes ended up on the cutting room floor, or maybe Sue Storm is actually blind. Yet, if she is blind, then why isn't she bumping into things?

Let's assume that she's really, really precise in her light-bending trick and only directs the tiniest bit of light directly into her pupils in a very controlled way. You'd still see two dark dots floating about, but maybe they'd be small enough that bystanders wouldn't notice.

Could this level of light-bending be done? I've already explained that another medium, like glass, can bend light – but, as far as we know, Sue isn't changing the medium she's moving in, or the properties of her own body. Although, in the 2005 film, *Fantastic Four*,[4] her clothes *didn't* automatically become invisible – making her more like H. G. Well's Invisible Man, who famously walks around swathed in bandages to disguise his condition. This suggests that her invisibility was somehow constrained to her body.

However, in the original comics, her clothes do disappear too[5,6] and this suggests that, whatever is happening, it is taking place outside her own body. Marvel tells us[1] that Susan Storm bends all wavelengths of light around herself with an act of concentration, so let's assume this to be true. Is it possible, then, to bend light with some sort of energy field?

Here, we run headlong into some rather unyielding physical problems. Light is composed of an oscillating electric and magnetic field, and there is a very important property of such fields: linearity. In other words, the two fields don't change each other at all. They just keep on going in their respective straight lines. You can't budge them with any kind of electromagnetic field, which makes sense if you think about it, else light would be constantly interfering with itself.

So, is there any force that can bend light? During the 1918 solar eclipse Sir Arthur Eddington famously proved[7] that the light reaching us from distant stars *was* being bent by our sun's gravitational field, thus confirming Albert Einstein's prediction from his general theory of relativity. Gravitational fields can, and do, bend light.

Yet, even our sun, which has a mass of approximately 2×10^{30} kg, only bends the light travelling to the earth from other stars by a tiny fraction of a degree. To make light pass around her body, Susan Storm must bend light much more dramatically than this. If she could actually generate the kind of gravitational field that would achieve this level of light-bending, she would, to quote one physicist I asked, 'disappear up her own black hole.' In short, if the Invisible Woman really did have the ability to bend light in this way, she'd only be able to do it once.

However, there are other routes to achieving invisibility, or at least the illusion of invisibility. In 2015, Xingjie Ni and Zi Jing published a paper in the journal *Science* titled 'An ultra-thin invisibility skin cloak for visible light.'[8,9] The 'cloak' is made of metamaterial: a material engineered to have unusual properties that are not found in nature. In this case it was made of an ultrathin layer of brick-like blocks of gold nanoantennas. The surface was engineered to reroute reflected light waves, so the object it was covering became effectively invisible when the cloak was activated. It's an important proof of concept, although the cloak itself was microscopic and only worked with a particular wavelength of red light – so it wouldn't effectively hide anything from human eyes (unless, perhaps, you had a really severely colour-blind villain).

Other works of fiction have proposed different ways to achieve invisibility. In the 2002 James Bond movie, *Die Another Day*, the Aston Martin Vanquish could become invisible at the touch of a button.[10] This works, we are told, by adaptive camouflage. This is real technology, although it will not have escaped readers' notice that invisible Aston Martins are not a regular feature on our roads. (Or are they?)

The idea of adaptive, or active, camouflage has a biological rather than technological origin. Animals have been doing this kind of thing for a very long time, particularly 'cold-blooded' creatures (exotherms) such as cephalopods, fish and reptiles, where it evolved not only to disguise animals from predators, but also to give them more control over their own body temperature – if you feel a bit chilly, you simply find a sunny spot and make yourself darker – and as a method of communication.[11]

Animals achieve adaptive camouflage in two ways: by producing light to blend in with a lit background or, more famously, by actually changing colour. For example, a tropical flounder can change colour to match different background textures in just a few seconds.[12]

Animals achieve this clever trick by using special chromatophore cells. These cells contain a pigment of one particular colour in a tiny, elastic sac surrounded by muscle and nerve cells. The sac is stretched or relaxed, which changes its opacity.[13] The overall effect is great for hiding against a gravel seabed, but isn't precise or quick enough to work effectively in a more visually complicated environment.

Nevertheless, the idea has been adopted in fiction. For instance, the villains in the Predator films famously make use of adaptive camouflage, as does at least one other superhero – Anole, of the X-Men, is said to be able to 'copy the pattern of any surface he stands against.'[14]

Scientists and engineers have also given the idea a lot of consideration, not least because of its potential military applications. Perhaps most famously, in 2003 researchers at the University of Tokyo created a material impregnated with retroreflective glass beads. A video camera behind the cloth captured the background, and then the image was projected onto a glass plate acting as a partial mirror so that some of the light was reflected onto the cloth and, ultimately, back to the viewer. The final effect was rather ghostly – the cloaked figure was still visible, but partially transparent.[15]

Unfortunately, it only worked from a certain angle. Getting this concept – of projecting the background image towards the viewer so they don't 'see' the object in between – to work on moving objects like cars from all viewing angles is still some way off. But it is far from impossible. Those Klingon cloaking devices could definitely be on their way.

However, there might be a much simpler way for humans to achieve invisibility without very much effort. In fact, people already do these kinds of things every day. There's a very famous experiment involving people passing basketballs around.[16] If you've never seen it, do a quick search for 'selective attention test' and watch it now. It works much better if you don't know what's about to happen ....

Done? Did you see the gorilla? If you've never seen the video before, the chances are you didn't. This is because of selective attention. We spend a lot of time ignoring huge chunks of what's going on around us because, if we didn't, our brains would quickly become overwhelmed with information. In particular, if we're focused on completing a certain task, we often 'tune out' to the rest of our surroundings. Magicians have been using these tricks for thousands of years and they call it 'misdirection'. For instance, they might get an audience to look in one place and, when they do, they easily miss something happening somewhere else, which would reveal the secret of the trick, were it noticed.

Perhaps it is possible to achieve invisibility by simply persuading onlookers that there's nothing there? This idea has been explored in the world of superheroes too. The Shadow is one of the oldest characters, first appearing in a series of pulp novels in the 1930s before making his debut in a radio drama in 1937.[17]

In the radio drama he was a mysterious avenger, described as a master of hypnotism who had 'learned the mysterious power to cloud men's minds, so they could not see him.'

He's far from the only fictional character to make use of such tricks. Terry Pratchett's witches, from the Discworld series of books, are famous users of 'headology' and have the ability to hide in plain sight. For example in *Maskerade*, Granny Weatherwax uses this skill in a scene where Pratchett writes: 'The old witch faded. She didn't disappear. She merely slid into the background.'[18]

This sort of approach certainly has its appeal: no need for huge amounts of technology – or its associated power supplies – no need to mess about with refractive indexes and blood bleaching (yikes), and certainly no need to create black holes in order to bend light. You simply manipulate people's minds instead, which is much easier.

But, maybe there's some other way that we just don't understand yet. According to the Marvel Database,[1] 'Mr Fantastic suspects that Sue Storm somehow taps into hyperspace when she uses her powers.' Perhaps she has some workaround for the laws of physics as we know them. She is a superhero, after all.

To quote Dr Ron Evans from a recent BBC Horizon documentary about anti-gravity and gravity control, perhaps 'It's not impossible. It's just that we don't know how to do it yet.'[19]

ACKNOWLEDGEMENTS

With thanks to the science teachers at the Royal Latin School in Buckingham for humouring me during several insightful chats about the biology and physics of invisibility.

REFERENCES

1. Susan Storm (Earth-616), *Marvel Database*, 2016 [cited 22 October 2016], available from: http://marvel.wikia.com/wiki/Susan_Storm_(Earth-616).
2. H. Wells, *The Invisible Man*, Atria Books, New York, 2014 [cited 22 October 2016].
3. *Disappearing Glass Rods*, Exploratorium, 2015 [cited 22 October 2016], available from: https://www.exploratorium.edu/snacks/disappearing-glass-rods.

4. *Invisible Woman*, En.wikipedia.org, 2016 [cited 22 October 2016], available from: https://en.wikipedia.org/wiki/Invisible_Woman#Film.
5. S. Lee and J. Kirby, *Fantastic Four #1*, Marvel Unlimited, New York, 1961.
6. *The Invisible Woman*, Nothingbutcomics.files.wordpress.com, 2016 [cited 23 October 2016], available from: https://nothingbutcomics.files.wordpress.com/2015/07/invisible-woman-kirby.jpg.
7. F. Dyson, A. Eddington and C. Davidson, A determination of the deflection of light by the sun's gravitational field, from observations made at the total eclipse of May 29, 1919, *Philos. Trans. R. Soc. London*, 1920, **220**(571–581), 291–333.
8. X. Ni, Z. Wong, M. Mrejen, Y. Wang and X. Zhang, An ultrathin invisibility skin cloak for visible light, *Science*, 2015, **349**(6254), 1310–1314.
9. *Making 3D Objects Disappear*, Berkeley Lab, News Center, 2015 [cited 23 October 2016], available from: http://newscenter.lbl.gov/2015/09/17/making-3d-objects-disappear/.
10. *List of James Bond vehicles*, En.wikipedia.org, 2016 [cited 23 October 2016], available from: https://en.wikipedia.org/wiki/List_of_James_Bond_vehicles#Aston_Martin.
11. D. Stuart-Fox and A. Moussalli, Camouflage, communication and thermoregulation: lessons from colour changing organisms, *Philos. Trans. R. Soc., B*, 2009, **364**(1516), 463–470.
12. V. Ramachandran, C. Tyler, R. Gregory, D. Rogers-Ramachandran, S. Duensing and C. Pillsbury, *et al.*, Rapid adaptive camouflage in tropical flounders, *Nature*, 1996, **379**(6568), 815–818.
13. R. Cloney and E. Florey, Ultrastructure of cephalopod chromatophore organs, *Z. Zellforsch. Mikrosk. Anat.*, 1968, **89**(2), 250–280.
14. Victor Borkowski (Earth-616), *Marvel Database*, 2016 [cited 23 October 2016], available from: http://marvel.wikia.com/wiki/Victor_Borkowski_(Earth-616).
15. *What is Optical Camouflage?* Innovateus.net, 2016 [cited 23 October 2016], available from: http://www.innovateus.net/science/what-optical-camouflage.
16. *Selective attention test*, YouTube, 2016 [cited 23 October 2016], available from: https://www.youtube.com/watch?v=vJG698U2Mvo.

17. *The Shadow*, En.m.wikipedia.org, 2016 [cited 23 October 2016], available from: https://en.m.wikipedia.org/wiki/The_Shadow.
18. T. Pratchett, *Maskerade*, HarperCollins e-Books, United Kingdom, 2014.
19. Horizon, *Project Greenglow*, BBC2: BBC, 2016.

CHAPTER 10

The Flash: The Fastest Man on Fire

BRIAN MACKENWELLS

Wellcome Trust Centre for Human Genetics, Roosevelt Drive,
Headington, Oxford, OX3 7BN, UK
E-mail: brian@mackenwells.com

10.1 JUST HOW FAST IS THE FLASH?

Barry Allen, forensic scientist at Central City, is tidying up his lab one evening before leaving work. He places the chemicals on a shelf, more-or-less at random. But then, out of nowhere, a bolt of lightning streaks through the window, hitting Barry and his shelf of chemicals. He wakes up to discover that he can connect to the 'Speed Force' – a mystical force that allows him to be the fastest man alive! That, by the way, is a good reason to study chemistry – you might get superpowers.[†]

Having become The Flash, Allen can now run at incredible speeds, which is a pretty cool superpower. Imagine it, you'd never be late! Or, more likely, you'd put off leaving the house for so long that you'd still be late; even superheroes like a lie-in. Yet, the Flash actually has

---

[†]You probably won't, though. Being hit by lightning and covered in chemicals is probably quite hazardous to your health.

The Secret Science of Superheroes
Edited by Mark Lorch and Andy Miah
© The Royal Society of Chemistry, 2017
Published by the Royal Society of Chemistry, www.rsc.org

141

another, secret, secondary power that makes his super-speed possible. So, let's look closely and see if we can work out what this secret power is and so describe the science that explains this capacity.

The first thing we need to know is how fast he runs. For this, we need to turn to the history books. In 1967 Superman and the Flash were asked by the UN Secretary General to have a race for charity.[1] For this, we can assume that the Flash was running as fast as he could – if you had the chance to beat Superman, you would, wouldn't you? He can run fast enough to skip across the ocean like a big red stone, so the race is all the way around the earth. To quote from a panel – 'And so, within seconds, the vizier of velocity and the super-sultan of speed approach the coast of Africa at 140 000 miles an hour...' Leaving aside, for a moment, that Africa is a continent and not a country, this panel gives us an actual number we can use for the Flash's top speed – 140 000 miles per hour. For those who prefer their units in metric, this is 225 308 kilometres per hour or, for any sailors reading, 121 656 knots. At that speed, you could run to the moon in a little over an hour and a half (assuming someone paved the way).

However, as the Flash pootles about Central City, saving people and fighting rogues, he's not running in a vacuum. He's running through air and air gets in the way. For the kinds of speeds we're used to in our day to day life,[‡] the fact that we're pushing our way through a big sea of air isn't usually a problem, or even something we particularly notice.[§] However, for the Flash, you start to get into unusual effects quite quickly.

10.2 CRASHING AND BURNING

Let's imagine that we are looking at the Flash really closely in super-slow motion. As he starts to run, he hits the air molecules in front of him. Those molecules – the unfortunate front ones – get pushed into the air molecules behind them, then they push the ones behind them, and so on. This 'push' backwards moves through the crowd of air molecules at about the speed of sound and it guides the air molecules which are further back in the crowd around the Flash. As he gets faster and faster, these waves of 'pushes' get closer and closer together. Eventually, when the Flash

---

[‡]Unless you're a fighter pilot, or an astronaut.
[§]Ed: I notice! I notice every day, when I cycle into the head winds on my way to work.

hits the speed of sound, he'll actually catch up to these waves and everyone he ran past on the street would hear a sonic boom.

This is where things really start to heat up, literally. As the Flash forces his way through the air, he's pushing all those air molecules together, increasing the pressure on them. Furthermore, as pressure increases, temperature increases; squeeze air enough and it gets really hot! The other thing to remember is that each time a little air molecule smacks against the Flash, it gives him a little bit of its energy. If you hit a piece of metal with a hammer for a few minutes, you'll notice that your metal is getting hot – it's the same process, but on a larger scale. The energy coming from your muscles, as you deliver the strike into the metal, must go somewhere and so it turns into heat. As the Flash runs, the energy from those tiny impacts need to go somewhere, and they turn into heat, amongst other things.

At this point, the physics of what is happening to the Flash starts to get extremely complicated. Each air molecule bounces back, hitting the air behind it fast enough that the Flash would have a little barrier in front of him, where the oncoming air only smacks against the air immediately in front of him, and never actually hits the Flash. This barrier of air in front of him is called the shock layer. However, as he gets faster and faster, the air bouncing around in this shock layer gets hotter and hotter, due to the pressure increase and these little impacts. Eventually there is enough heat in this shock layer to rip the air molecules apart. This releases all kinds of light and heat. So, if you go really, really fast, you start to get really, really hot.

It's actually pretty easy to see an example of this in the world around us. In fact, it happens anything between 4 and 16 times an hour, depending on where in the world you are, the date, and what time it is. All you have to do is find somewhere remote, with a nice view of the night's sky – then you can see the meteors as they burn through our atmosphere, compressing the air in front of them just like the Flash does. So, why isn't the Flash a human, flaming meteor, catching fire as soon as he starts to run at super-speed? How does he protect himself from the heat?

10.3   BEAT THE HEAT

Engineers have needed to figure this out already, as it's the same problem that confronts space explorers – in particular getting astronauts back into our atmosphere after being in space. To do this, in very rough terms, you need to throw them back at the Earth at roughly 17 000 miles per hour[2] and stop them burning up as they tear through our atmosphere. Quite a lot of thought has gone into how to safely return fragile, fleshy, human beings back to the surface without them ending up looking like a burnt pizza. It was very important during the early days of the space programs to understand the physics of re-entry. Back then – during the cold war – the US (and, presumably, the Russian government) did a lot of research into how their intercontinental ballistic missiles are affected by their speed. There is now a feel for how much shielding is needed to soak up enough heat to protect your astronauts or missiles, so there hasn't been much work done on it since. This means that, to work out exactly how much heat the Flash produces by running at 140 000 miles per hour, declassified US government reports from the cold war have been particularly useful. That, and fans of Kerbal Space Program who have been trying to work out how to realistically model re-entry temperatures in the game.

So, after all that, we can use the 'isentropic total gas equation' to figure it out, partly because it is what NASA use, and partly because it is relatively simple. It makes some assumptions about how air acts when hit by someone running at Mach 182, but it should give us a rough idea. Plugging in the various numbers gives us a temperature of around 3.4 million degrees Celsius, which is about the same temperature as the Sun's corona. Strictly speaking, this is called the 'stagnation temperature' – the temperature just outside the shock layer, because (slightly counter-intuitively) the shock layer adds a kind of buffer where the molecules aren't moving as quickly. There are a few reasons for this, but the main one is that it takes energy for heat to conduct into the Flash – the energy to do this comes from this shock layer. So, the actual heat of the layer closest to the Flash's skin will be a lot lower than the stagnation temperature. However, we can observe that the air around him doesn't heat up at all, as he doesn't burn Central City to the ground every time he runs, which suggests that he's soaking up the full brunt of the temperature increase around him. How does he achieve that? There are three strategies from space re-entry module design that are instructive here.

The first is called 'ablation' and it uses shielding on the leading edge of the re-entry module that is designed to melt or vaporize entirely. The moment that some of this shielding changes from a solid to a liquid or gas, it soaks up heat from the re-entry to fuel that change. The Flash doesn't have any parts of his costume, or skin, that melts off as he runs, so that's probably not the strategy he uses.

The second strategy engineers use is 'emission'. This is what happens when you heat something up so hot that it glows – it takes energy to give off light and heat like that. If you design your shielding right, you can make it emit some of the heat it has absorbed. That is why space shuttles have very dark tiles on their underside, which have been specially made from silica to emit a lot of radiation, like light and heat. However, the Flash doesn't seem to glow red hot when he's running, so emissions can't be what he uses to lose heat.

The third strategy is the oldest and the best candidate for explaining what's going on with the Flash. When engineers were designing intercontinental ballistic missiles in the 1950s, they packed the nose with material that would just soak up and store the heat – called a heat sink. If you imagine a big pot of water and a small pot of water, each over a flame of the same size, which would boil first? The smaller one would, because less water is storing the same amount of heat, so the peak temperature in the smaller pot gets higher quicker than the water in the bigger pot. Thus, the more heat-sink shielding you put into your rocket, the more heat from re-entry will get soaked up, and the peak temperature of your re-entry module will be lower. However, there's a trade-off when you're designing a space-craft – the heavier it is, the more rocket fuel you need to get it into space in the first place. You then need extra rocket fuel to lift the extra rocket fuel and you need more fuel to lift that fuel. The ratio quickly becomes untenable, so they don't use heat sinks very often any more, as the other forms of shielding work well and are lighter. However, it's the most likely scenario for why the Flash can run so fast and not be the fastest man on fire.

The only obvious heat-sink that the Flash has is himself, since he doesn't carry around a huge piece of shielding with him at all times. As it happens, people are mostly water, which is a fantastic heat-sink – it's why, if you have a smart-meter on your energy supply, you can see your energy usage shoot up when you boil the kettle. It takes a tremendous amount of energy to heat water up to boiling temperature.

How much energy, or heat, a material can soak up like this is called its specific heat capacity (SHC). The higher the SHC, the more energy it can absorb. Water has a SHC of 4.187 J g^{-1} K^{-1}, which means that it takes 4.187 joules of energy to raise one gram of water by one Kelvin. But the Flash can apparently absorb a lot more energy than water. So what's his SHC?

The equation we need to solve to establish the SHC of a material is not too complicated – you divide the amount of energy by the mass, multiplied by the range in temperature. Since he's absorbing 3.4 million degrees and it doesn't raise his body temperature by much, and according to the internet he's around 90 kg in weight, we can work out that he has a specific heat capacity of a little over 7 billion $J\,g^{-1}\,K^{-1}$. Remember, water has a particularly high SHC of around 4. So, this would seem to be the Flash's secret secondary superpower: the human heat-sink. You probably couldn't use it to fight any crimes by itself, but it pairs very nicely with super-speed.

You are now in possession of highly privileged information, because we can succeed where a host of rogues and time travellers have failed. We can defeat the Flash. Why would we want to do that? Well, who knows. Maybe you're a time traveller from the 25th century and you try to commit some crimes only to be constantly foiled by the Flash. Maybe you're a hard-working blue-collar super-criminal with freeze powers, and you just want to be able to knock over a bank in peace. To defeat the Flash, we just need to think of a situation where you would want a higher body temperature, since that's what he would struggle with. Luckily this is an incredibly common situation and it has almost certainly happened to you in the last year, probably in the winter. We need to give the Flash a cold.

When you contract a cold, you suffer from a host of symptoms. You sneeze, you cough, you realize that there isn't actually much on telly during the day. However, the symptom that matters most to us when fighting the Flash is the fever that you will experience. We know that people are more likely to get colds during the winter, but we are still not sure why. One theory is that, at lower temperatures, the walls of the virus might be a bit harder, so your body can't fight it as effectively. Alternatively, it might be that cold weather lowers our immune system, making us more susceptible to illness. Or, perhaps it is just because we tend to huddle indoors in groups more in the winter than during the summer, making it easier for viruses to spread. It could be a combination of all of these.

What we do know is that, when you catch a cold, a part of your brain called the hypothalamus starts to release hormones that tell your body to start raising your temperature. This is pretty unpleasant for you, making you alternately throw off the covers and pull them up. But, the theory goes, it's even more unpleasant for the virus that's causing your cold – your body is trying to create an inhospitable environment for the virus, making its walls softer, and so easier for your immune system to fight against it.

In order to raise your body temperature, that energy needs to come from somewhere. It's one of the reasons you get really weak and tired when you have an infection – energy you'd normally be using to sit around and read comics, your body now needs to fight the invader.

Now, we have discovered that the Flash has a huge specific heat capacity. That means that to raise his body temperature by even one degree will require an enormous amount of energy. The only way we have to get energy into our systems is through food, and since there's never been a hint that the Flash can photosynthesize, he must need to do the same.

If we move around some of the terms in the equation we used to discover his SHC, we can work out how much energy the Flash would need to raise his body temperature. The exact temperature you need to reach in order to be clinically diagnosed as feverish depends on the location of your thermometer, but aiming for an increase of one degree should cover it. Plugging in those values to our equation, to raise the Flash's temperature by one degree we would need 651 billion joules. Now, the joule is a great unit of measurement. A personal favourite, honestly. The joule is part of the SI, or *le Système international d'unités*, system of measurement, used by scientists all over the world. However, 651 billion joules is a bit difficult to imagine. So, for our purposes, a far more sensible unit of measurement would be the 'tin of chicken soup'.

A tin of cream of chicken soup has (according to the back of this tin from my local shop) 142 Calories. A calorie is the amount of energy you need to raise one gram of water by one degree, but the Calorie listed on my tin is subtly different to this calorie. The tricky bit here is the difference between the calorie and the Calorie. Pay careful attention to the capitalization there. You see, if you count up the number of calories (*i.e.* amount of energy) in our tin of soup, you get 142 000 calories. But the back of your food packet actually lists the number of *kilocalories* that are in your food. The food manufacturers aren't trying to trick you – if you're watching what you eat, and counting calories, you're actually counting kilocalories, but since not much food has fewer than 1000 calories they've just dropped the 'kilo' bit. To make this difference clear, when you're talking about food calories (*i.e.* kilocalories) you capitalize 'Calories', and when you're talking about the other kind of calorie (the one related to heating up a gram of water) it goes uncapitalized. If you look carefully, on the back of food packets it should always be capitalised. I'm not convinced this capitalization plan actually makes anything clearer, but that's the sort of thing *le Système international d'unités*, was set up to sort out.

So, our tin of soup has 142 Calories – this is equal to 142 000 calories, which comes in at just under 600 000 joules. Consequently, to feed his fever of 651 billion joules, the Flash would need to eat just under 5 million tins of soup and that's just to get his temperature *up* to a fever – his body would need to keep that up until the virus had been dealt with. You need to time your sneeze very carefully, but if you can catch the Flash with it you will either lay him low with a *terrible* cold, or he'll bankrupt himself trying to keep himself fed. Success either way!

However, the Flash probably has a tertiary superpower, which we have yet to consider. You see, one of the rules of the universe is that you can't destroy energy – it has to *go* somewhere. The Flash runs around all day, soaking up that energy from the constant re-entry procedure he's undergoing – it has to go somewhere. Since he does not (as far as we know) create the loudest farts known to humanity, he must be able to convert that energy into a form his muscles can use, because running around so fast would use up a huge amount of energy. As such, he converts the energy from this heat directly into energy his legs can use to run really fast. He doesn't need to eat – he consumes heat! So to defeat a virus, he just needs to run around, which is how he fights baddies anyway.

However, our careful examination of his powers presents one final advantage to us. Since he consumes heat from his surroundings, on a really hot day, all you need to do is stand beside the Flash – it should be the coolest spot for miles. Now that's practical advice you can use!

REFERENCES

1. J. Shooter, C. Swan, G. Klein, M. Weisinger and E. Nelson Bridwell, *Superman*, vol. 1, #199 DC Comics Inc., New York, 1967.
2. NASA, *Entry, Splash Down and Recovery*, 2016 [Cited November 11], available from: http://history.nasa.gov/SP-4029/Apollo_18-40_Entry_Splashdown_and_Recovery.htm.

You've Got to Learn to Be More Flexible: The Mechanics of Marvellous

DAVID JESSON

Department of Mechanical Engineering Sciences, University of Surrey, Guildford, Surrey, GU2 7XH, UK
E-mail: d.jesson@surrey.ac.uk

11.1 INTRODUCTION

When thinking about the world of superheroes, a perennial question is 'what superpower would you choose to have?'. It's a question that is not easily answered well, especially as one digs deeper into the consequences of having certain powers. Indeed, the familiar adage – with great power comes great responsibility – has found roots even within writing about superheroes. In this case, the quote is found in Spider-Man, when Peter Parker recalls advice he was given by his Uncle Ben (*Amazing Fantasy*, #15, August 1962). One of the anxieties about having too much power is the uncertainty around the side-effects of power. For example, while invisibility is often a popular choice of superpower, it would

The Secret Science of Superheroes
Edited by Mark Lorch and Andy Miah
© The Royal Society of Chemistry, 2017
Published by the Royal Society of Chemistry, www.rsc.org

wreak havoc on attempts to ensue rights to privacy in a world that is increasingly concerned about the rise of the surveillance state.

On a personal note, I grew up with the 1981 Spider-Man cartoon television series, which aired together with a parallel series, *Spider-Man and his Amazing Friends*. While I like other heroes, Spider-Man seemed a bit more human than many others, and being a web-slinger seemed like quite a good bet[†].

But if you get the choice of superpower, then side-effects are definitely something that need some thought – 900 eggs for breakfast, for example, seems like a pretty high price to pay.

In this chapter we're going to explore "the mechanics of marvellous" – the consequences and limitations (and one unconsidered side effect) of someone whose ability to stretch might seem...incredible...

11.2   NATURALLY FLEXIBLE

If you take a look at Chapter 2, you'll see some examples of naturally occurring 'superpowers' and how these might evolve into something worthy of a crime fighting, world saving hero. In terms of flexibility, there is already some magnificent stretchiness in the animal kingdom. Small rodents like mice can get through what seem to be impossibly small holes. For them, the fundamental limitation is the size of their skull, but even here we don't know exactly what this limit is. The width of a biro is often mentioned (approx. 8 mm, probably applies to juveniles) while the size of a UK penny (20.3 mm) is a good approximation for the average adult. Both these figures seem remarkable when considering the visual appearance of a mouse. These are creatures with a very flexible skeleton, loose joints and stretchy ligaments. Cats are similarly lithe, and can fit through smaller gaps than one might expect, using their whiskers to sense their way through places.

[†]If you've seen Cool Hand Luke, then you might be thinking that 900 eggs for breakfast is a bit of a tall order. Obviously other protein sources exist, but let's stick with the eggs for a second. In the film, Cool Hand Luke eats 50 eggs in under an hour. This is a man who has not trained for the feat, so we'll be genuinely impressed. It is the kind of thing that makes people go "is that really possible?". Debra Ronca[1] on *How Stuff Works: Science* has a good analysis and shown that not only is it possible, but if you are a competitive eater in training, you can do considerably better than 50 in an hour – apparently the world record is 141 in 8 minutes, so Spidey could get his quota eaten in under an hour. Of course this is one of those side-effects which has side-effects of its own...

If you want to get really stretchy though, you need to lose your skeleton completely. For instance, octopi have no skeleton and, as a result, can fit through extremely small holes. National Geographic have shown how a 600 lb (approx. 272 kg) octopus can squeeze through a hole that is about the size of a US quarter coin (24.26 mm).[2] To do this, octopi make use of something called muscular hydrostatics, which means that, whilst the volume of a given octopus remains constant, the shape can be changed considerably.

This muscular hydrostatics capacity is shared among other species too, including humans. Each of us has a hydrostatic system in the form of our tongues, which can change shape in the same way as the octopus. In terms of superheroics though, your tongue is not the best example in the animal kingdom. For instance, some species of frog have tongues that can extend by 180%.[3] In other words, at full extension the frog's tongue is nearly three times the length of the tongue at rest. We need to bear in mind that in both these examples, we are dealing with something that is highly flexible, relatively powerful, but which has no structural rigidity – there is no skeletal structure. It is also worth noting that muscular hydrostats are predominantly water based, and therefore highly incompressible. The change in shape is effected solely by the expansion and contraction of muscle tissue.

Elastigirl – also known as Mrs Incredible – is the superhero identity of Helen Parr (neè Truax); she has the powers of super-flexibility and shape changing. Elastigirl also has some degree of invulnerability and superhuman strength (although nowhere near the superstrength of Mr Incredible). She also happens to be an excellent pilot, tactician and homemaker.

Whilst this chapter is much more about the mechanics of her superpowers (and, of course, the chemistry that supports these), it is also worth noting some of the psychological aspects that underpin her character. This tells us a lot about the relationship between any given superpower and what it takes to use it well. For example, Elastigirl is extremely driven and determined to stand out as a top hero in what she perceives (probably correctly) as a male-dominated world. She is authoritarian: she is the one that holds the family together (whilst husband, Mr Incredible, is depressed about having to live in hiding) and is the *de facto* lawmaker and arbiter. When her children Dash and Violet argue, or when Dash breaks the rules, it is Helen who lays down the law.

While some of the claims vary, Pixar states that Helen is 167 cm tall (just under 5 foot 6 inches), and that she can stretch any part of her body up to 30 m in length.[4] Further, it also states that she can reduce a body part to a minimum thickness of 1 mm. This conflicts with the Operation Kronos file on her, which states an extension of 300 m, but we can probably assume that the information came from Syndrome's recollection from before she went into hiding; Syndrome doesn't seem like the most reliable of witnesses.[‡]

There are a few things we can deduce from this. Hypothesis #1: she has no real skeleton, probably nothing beyond a skull, some sort of vestigial spinal column and possibly a pelvis. I'm assuming something close to a true skull, because Helen is still human, and not a squid. Mind you, we don't know her origin story... Hypothesis #2: she does have a skeleton of some description, but one that lacks the inorganic components (calcium hydroxyapatite and osteocalcium phosphate). These provide most of the compressive strength of bone, so there would need to be something replacing them. Likewise, collagen does not have the properties required to be stretchy enough for someone to extend their arm to 30 m in length.

11.3   MECHANICAL CHARACTERISATION 101

Scientists like hypotheses – it's the first step to understanding what's going on. But a hypothesis is just so many words if you don't test it. So firstly, a quick sanity check: if we make the very basic assumption that a limb is a cylinder, then it will have a volume of $\pi r^2 l$. As a rule of thumb, arm-span is nominally equal to height – try it with your own arms. Elastigirl has quite a slim build, so her finger-tip to arm is probably about 61 cm, and her shoulders are approximately 46 cm across. Hands are typically 10% of the length of the arm, *i.e.* ~6 cm, so that gives us an arm length of 55 cm. Again, very crudely, if we imagine an arm diameter of about 7.5 cm then this would give us an arm volume of 2430 cm³. With an average density of the human body being

---

[‡]Spoiler alert...If you've not seen the film, Syndrome is the villain of the piece, and has some knowledge of the supers from before they are forced into hiding. His master plan is Operation Kronos.

0.985 g cm^{-3}, this would give us a mass of 2.39 kg. For the average woman, an arm should weigh between 4% and 5% of their total body mass, so, for Elastigirl, assuming normal rules apply, we're looking at about 2.5–2.8 kg. However, when we factor in the hand and that we are assuming less bone than normal, then this figure looks pretty good. So far, so good.

Now, about that 30 m reach. Starting with the assumption that volume will remain the same, as with our muscular hydrostats, on that basis we can start with the volume that we've already calculated. This then gives us a radius of 5 mm or a diameter of 1 cm at full arm stretch. That's an order of magnitude greater than the 1 mm stated above, so we do need to refine our assumptions a little. There is probably more of a skeleton than we might have initially thought, and this is the limitation of her stretch. Localized thinning is possible, but must work around the skeletal structure.

Great. So we've put some limits on what is going on. Now we need to have a think about how Elastigirl manages her extraordinary stretch. Although, before we get down to the detail of all of that, we need to have a frame of reference. In the mechanical characterization of materials there are a number of fundamental properties that interest us. Here we only need to worry about three of them: stress (σ), strain (ε) and stiffness (E). In common parlance, stress and strain are often treated as being interchangeable terms for the mental state of having too much to do and not enough time in which to do it in (like trying to write a book in a weekend). However, from an engineering perspective they mean very different things. (Stress, strain, Young's modulus and the like are also covered in Chapter 7, about vibranium).

Let me explain: if I take a rubber band and pull on it, then I am applying a force to it by pulling on it. Stress is how hard I pull divided by how thick the band is and the units we use for this are pascals, Pa. Meanwhile the strength of the band is the stress required to break the band. Most materials have stresses measured in megapascals, MPa, *i.e.* millions of pascals.

Strain is very different from stress. It is basically a measure of how far the something (say the rubber band again) will stretch before it breaks. More specifically strain is the ratio of how much the band stretched divided by how long it can get before snapping. If the rubber band starts with a length of 10 cm and we apply a load and it stretches by 20 cm then the strain is 20/10 or 200%.[§] Now compare that to a 10 cm long bit of string; it might only stretch by just 1 mm so its strain is just 1/100 or 1%.

Stiffness is slightly more complex, but only very slightly. Back in the 17th century, Robert Hooke stated '*ut tensio, sic vis*', which means 'as the extension, so the force'.[5] He observed that elastic materials followed a law by which the amount they can be extended was proportional to the force applied. Different materials extend by different amounts for a given applied force. Think of this like pulling on the bit of string and then the rubber band. You can apply the same force to both, but one will extend more than the other. Now here's the important bit – Hooke found that however much a material stretches, the stretch will double if you double the applied force, triple if you pull it three times harder, and so on. Of course there is a limit, and things go PING, which we'll come onto in a moment. Hooke was a renowned polymath, and this perhaps explains why he got distracted and so didn't take this to its logical conclusion, *i.e.* that there is a relationship between stress and strain.

Back to the string and rubber bands; they both might have the same strength (say 10 MPa), but the string's strain is 1% whilst the rubber band's strain is 200%. Stiffness, then, is the strength divided by the strain. This relationship is usually called the Young's modulus, E, of a material: $E = \sigma/\varepsilon$.[¶] There is a very slight wrinkle, which we do need to be clear on. E, stiffness, can be used with other materials where slightly different behaviour is observed: the term 'Young's modulus' is reserved for linear-elastic materials only. It should also be noted that stiffness is sometimes used to denote a resistance to force applied: stiffness is treated as being the opposite or inverse of compliance. It can all get slightly confusing, even when you are using these terms every day – so it's probably best just to ignore it for now.

The other thing that we need to understand when we're characterizing materials is what we mean by *elastic* and, for that matter,

---

[§]Typically the change in length is labelled Δl. By dividing the change in length by the original length ($\Delta l/l$) we get the strain, which is usually given the symbol epsilon (ε). The capital Greek letter delta (Δ) is generally used to denote the change in something.

[¶]As in Thomas Young, another polymath, who in addition to many other feats helped to translate Egyptian Hieroglyphs using the Rosetta Stone. His paper on the subject of E dates to the very early 19th Century.

plastic as well. Elastic and plastic are two words that also have to do heavy duty in the English language. It gets especially complicated when we talk about the elasticity of plastics! Scientists have a bit of a reputation for being obsessed with classifying things. This is because what we would really like to do is predict behaviour from shared traits, but to achieve this we need to have some sort of understanding of what those traits are. So in materials we talk about metals (for example adamantium), ceramics, plastics and composites (possibly including vibranium – see Chapter 7). Plastics, as a class, are long chain hydrocarbons, with various things added (functional groups) or taken away (to give unsaturation, *i.e.* double and triple bonds).

Plasticity, in terms of materials behaviour, is the name given to deformation that is non-recoverable. The difference is that an elastic material is able to deform AND recover its shape. If you are struggling with the difference between plastic and elastic then making a pizza might help. Start with the dough. If you knead it the dough deforms but springs back towards its original shape, like an 'elastic' band. Next make the topping and be sure to add loads of mozzarella cheese. Before you finish eating your pizza, can I ask you to take a bite, pull and observe ... Note how the cheese stretches between the chunk in your mouth and the rest of the slice in your hand. But the mozzarella doesn't regain its shape because it is plastic.

So what's going on with all this stretching and straining at an atomic level? In metals we can stretch a sample and recover the properties if we refrain from actually moving the constituent atoms in relation to each other; we just pull them apart slightly. Similarly, plastics can be stretched – some quite considerably – without being permanently deformed. Plastics are, perhaps, more prone to plasticity, but most materials demonstrate this behaviour to some extent.

Now, confusingly, things can be both elastic and plastic (and sometimes fantastic). We can stretch a material, it deforms elastically until it reaches a key point, and then it deforms plastically. The point at which this change in behaviour occurs is called the yield stress. Most of the time, engineers try and keep below this value, although there are some applications where the material is deliberately allowed to yield: the change in microstructure and the movement of defects can give rise to some interesting properties. In terms of strain, some materials, such as polypropylene can extend considerably post-yield. The force that needs to be applied to produce this continued extension actually decreases, because the material 'draws down' and becomes thinner, so the stress that is applied is relatively constant.

11.4 HOW TO AVOID WHIPLASH IN YOUR (SUPER) 'QUICK CHANGE ACT'

Clearly, whatever materials are underpinning Elastigirl's abilities (and for that matter her suit, created by Edna Mode), she needs to be working within the elastic limit of these materials (or, the psuedoelastic limit; again, see Chapter 7).

This of course begs several questions! Taking, a quick detour, there is another factor that we need to consider – the rate at which this flexibility occurs. Generally, there are three rate regimes that interest us: quasi-static, dynamic and ballistic. Quasi-static is a slightly fancy term used in mechanical characterization of materials to mean 'extending the test sample very slowly'. We're typically talking about a few mm per minute at most. Ballistic speeds start at around 7 200 000 mm per minute (*i.e.* around six orders of magnitude faster than quasi-static speeds). Dynamic, in a rather woolly way, fills the gap in-between.

Some materials that are very stretchy under relatively normal loading can behave in an almost brittle way when loaded very quickly. Others, whilst they might not break catastrophically, can exceed their safety ratings. For example, a large proportion of modern climbing ropes are manufactured from nylon, which has excellent properties for this kind of application. Nylon has a good specific strength (*i.e.* it is strong but light), flexible but not overly so, and can be coiled without damage. It is a popular choice of material for most climbers. It is also used for abseiling, which – scientifically speaking – it is tempting to describe as anti-climbing. Abseiling always looks like great fun, until you are actually walking over the side of the cliff. Typical scenes in movies involving this sort of behaviour will usually involve the hero jumping down the side of a 1000 m cliff in three bounds, or elite soldiers leaping out of a helicopter and stopping just before they hit the ground. In both cases, the effect of the stress on the rope is exacerbated by the shock on it, as the load is applied suddenly.

To illustrate this, a study by Harutyunyan *et al.* (2016) examines the ideal dynamic properties of climbing ropes. The shape-memory effect was identified as a possible mechanism for allowing ropes to provide constant deceleration rather than instantly stopping. Unfortunately, the materials that could deliver such a shape-memory effect do not yet exist, but identifying them has benefits for the experience of climbing, and also for the durability of the ropes.[6] If we relate this to Elastigirl's abilities, it suggests a mechanism that would enable her to avoid 'whiplash' in her arms and legs whenever she changes shape quickly.

This ability for her body to react quickly also gives us a clue to her invulnerability – she is her own body armour, using energy absorption to deal with most problems. In real life, different methods need to be used to deal with different threats: even assuming that two different threats involved the same kinetic energy, the shape of the impactor would have an effect on the overall result. The corollary here is the factoid of a woman in stiletto heels exerting the same force as an elephant. Because the mass is concentrated to a small cross-sectional area, the force applied on that area is large. On the other hand – or perhaps foot – the elephant's weight is spread out over four relatively large disks. The combination of the total energy, the area over which it acts and the rate at which it acts means that you see very different behaviours. Many threats can probably be avoided by dodging them completely; others will see autonomic flexing of the body to change shape and absorb the energy, shedding it over a larger area.

11.5 THE MECHANICS OF MARVELOUS: PART 1 – JUST A MOMENT

Returning to the concept of how far Elastigirl can stretch, 30 m didn't seem unfeasible on the basis of a constant volume being extended out. What if we think of this in terms of the strain capacity (remembering that we're dealing with a material that we don't want to extend beyond its elastic limit, or yield stress)? On this basis, then, we need to calculate the strain that is being applied, which comes out at about $\varepsilon = 54$. Looking at Figure 11.1, that isn't impossible for some of the types of materials that we might be

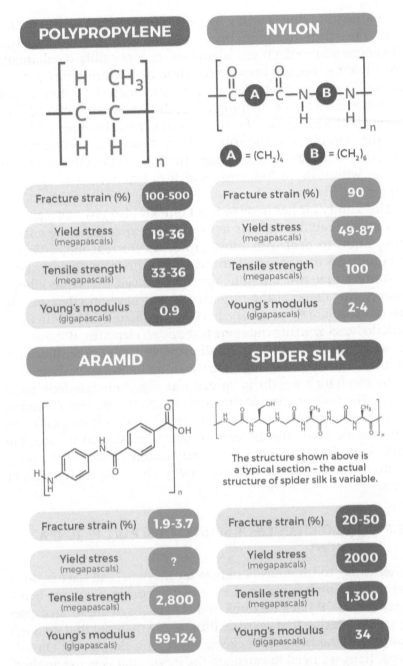

Figure 11.1 The chemical structure and some mechanical properties[7] of known materials which may help to understand Elastigirl's superflexibility. © Andy Brunning 2017.

interested in, although we need to remember that these are properties of the material at failure – the strain at the point of yielding
is considerably lower. What is also interesting is that a material
that is known for its ability to absorb energy (aramid, a form of
which is Kevlar®) is surprisingly brittle. There are mechanisms
for absorbing energy, but the material itself is not particularly
stretchy. Some forms of spider silk are much more appropriate in
this context, being polymeric shape-memory materials.

Now that we understand a bit about the stretchability of some
superheroes, it is worth considering the consequences of such
capacities. In engineering terms, one important concept is called
'moment'. This is not a fraction of time, but rather the consequence of load applied and the distance that application is from
some sort of fixed point. To illustrate this further, let's try a little
experiment in interactivity.

You'll find this experiment easier if you stand up. This isn't necessary, but some space around you is crucial. Next, extend one
hand away from your body (it is important that you extend the
arm that isn't holding this book or you might struggle to read the
rest of the instructions): make your arm straight, but do not over-
stretch. It doesn't matter whether you extend your arm out to the
side or out in front. Hold that for a few seconds: be mindful of
your arm, and then feel the weight of your fingers, and compare
that to the sensation of weight at your shoulder, which is where
the whole mass of your arm is hanging off your body. OK? Now
take a rest!

Now, find something that you can hold in your hand com-
fortably. Ideally something like a cup or mug (perhaps filled
with the beverage of your choice). If you don't have something
like that to hand, you could use your phone or your wallet. No,
don't use this book! You need to be able to keep reading. Right,
when you are ready, repeat the exercise, holding out the hand
with your chosen object. Hold the object out at arm's length
for a few seconds. Again, be mindful of your arm, feeling it
extended but not over-stretched. Starting from your fingertips, scan inwards to your shoulder and feel the sensation of
increasing 'weight'. When you get to your shoulder, you should
be able to feel a difference between holding and not holding
an object. What you're feeling is the increasing moment due
to adding a mass at a distance. As anyone who has ever had

training in heavy lifting knows, it is better to keep the mass close to your body. As well as being more stable and easier to handle, the key thing is that it is better for your back – you're reducing the effect of the mass on your body, by keeping the mass close to your centre of gravity and hence limiting strains on your body.

We can attach a numerical figure to the impact of this in each position. The moment is zero at your fingertips and increases at a constant rate as you get closer to the shoulder. If we were being very conscientious, then we would account for the weight of the arm itself in all of this, but we can just take the weight of the object, which is probably a few hundred grams (unless you've been very enthusiastic and used a large dictionary as your chosen object), and your arm length is probably of the order of 0.4–0.7 m. For the purposes of the calculations we're about to do let's split the difference and say arm length is 0.55 m.

What if Elastigirl is having a morning cup of coffee. The cup of coffee will have a mass of about 500 g (or 0.5 kg). Gravity is acting on the mug, which means the downward force of the mug is equal to the mass times acceleration due to gravity, which comes to 5 newtons (5 N). The moment (M) is equal to the load (which we've just calculated to be 5 N) multiplied by the length of the arm (0.55 m). This gives $M = -P \times L$.[||] Plug in the numbers and you get $M = -5 \times 0.55 = 2.75$ N m. That's not too bad, and is fairly easy to manage – after all you probably didn't struggle to hold your cup out at arm's length.

But what if Elastigirl has met up with friends at her favourite café... She sits down whilst she waits for the barista to brew her beverage. Then, not wanting to miss a moment (of time) of the conversation she extends out an arm to its full 30 meters (it is a large café) and grabs the mug. Now L is 30 m instead of 0.55 m, so the moment Elastigirl's shoulder now feels is over 50 times greater, or about 150 N m. This would be the equivalent of you doing the exercise we tried earlier whilst holding a bucket containing 27 litres of coffee (or a large sack of potatoes, if you

---

[||]The sign convention we use to describe a force moving in the direction of gravity is negative, hence $M = -PL$.

find volumes hard to imagine). Not impossible (for someone who spends plenty of time in the gym), but you would certainly need to pay attention to what you were doing (unless you're a superhero).

However, lifting a cup of coffee is not particularly heroic, no matter how long your arms. In action, Elastigirl is going to be lifting and throwing much larger objects. So, for example, if we consider her lifting a henchman in preparation for throwing him, then a 100 kg thug represents a force of 1000 N (1 kN). At normal arm-length distance, this works out at a moment of 550 N m (four sacks of potatoes now), whilst at full extension we get up to a whopping 30 000 N m. Again, if you wanted to feel the effect of this at home then you would need to try and lift an (African) elephant. (On my lawyer's advice, I'm required to say, do not try this at home).

11.6   THE MECHANICS OF MARVELLOUS: PART 2 – A MOMENT'S DEFLECTION

Yet, the moment that is created is only part of the story. Recalling the three properties that were mentioned earlier in the chapter, we also need to consider the strain that is generated by carrying the load, and what this might mean in terms of the stiffness that might be required for the material, or in this case the structure.

As alluded to at the beginning of the chapter, different materials have different properties and, to a great extent, these properties are dependent on the structure of the material. Look at Figure 11.1 and you'll notice that, even within the same class of materials, some very different properties emerge. For instance, some materials are relatively stiff and some are less so. There are also materials that have a high strain to failure and some that are rather brittle. A brittle material can carry large loads but fails after a small deformation. Consequently, thinking about the exercise that we tried earlier with your cup of coffee, the force applied will cause your arm to deflect slightly (though this would be relatively imperceptible).

To illustrate this further, imagine a diver at the end of a springboard. The stiffer the board, the smaller it will deflect in reaction to the load of the diver. Skipping through more equations** we encounter a problem. To heave a henchman, Elastigirl's arm needs to be stiffer than is physically possible …

The deflection that is caused by picking something up exceeds the capacity of Elastigirl's arm. On this basis, we can just about accept that she could pick up a cup of coffee from about 15 m away, but she would need to be careful not to spill any. She could lift a henchman off the floor, but he would have to be relatively close – although still out of normal arm range, which gives her some advantage. As a note, even if we were somehow able to increase the stiffness of the material by a factor of 10, the maths tells us that Elastigirl's range would only increase by a factor of 1.6. This is illustrated in Figure 11.2: our understanding of how atoms interact would have to change drastically in order to find anything considerably stiffer.

**If you are interested, the key equation is:

$$\frac{M}{I} = \frac{E}{R} = \frac{\sigma}{y}$$

Just to recap: M is the moment, E is the stiffness and σ is the (applied) stress. I is a property called the second moment of area, which is a function of the geometry of a structure. Let's stick with the simplified cylinder, which gives us:

$$I = \frac{\pi r^4}{4}$$

R is the radius of curvature of the bend in the beam and y is the (maximum) displacement due to the (applied) stress. With a bit of mathematical magic, we can derive a value for R in terms of the displacement:

$$R = \frac{L^2}{8y}$$

If we take all of this together then we can place E in terms of y, like so:

$$E = \frac{MR}{I} = \frac{PL \cdot \dfrac{L^2}{8y}}{\dfrac{\pi r^4}{4}} = \frac{PL^3}{2\pi r^4 y} = \frac{1}{y} \cdot \frac{PL^3}{2\pi r^4}$$

This means that E is proportional to $1/y$: this is the mathematical proof that, as the stiffness increases, the displacement that is created by a given force decreases.[8] Plugging in the numbers for the henchman and the radius of an arm and we get a value of stiffness which unfortunately exceeds that of a carbon–carbon bond (1×10^{12} Pa). And that is about as stiff as you could make anything.

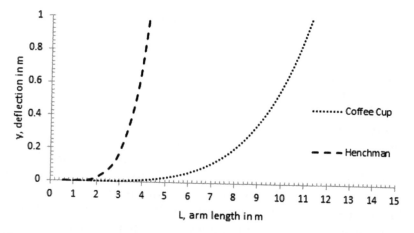

Figure 11.2 The change in deflection of Elastigirl's arm as a result of picking up a mass (a coffee cup, 0.5 kg, or a henchman, 100 kg) at different arm lengths.

11.7 CONCLUDING REMARKS

So, what have we learned from all of this? One thing that is really clear is that superhero science is not straightforward! Amongst many characters, there are some great examples of the known laws of physics being obeyed and others where they are flouted completely. We can also deduce that Syndrome from *The Incredibles* is not the super-scientist that he professes to be. We probably already knew this, but he has made a fundamental mistake somewhere along the line. Indeed, one thing that most superheroes and supervillains have in common is hubris: Syndrome believes his data is infallible, rather than subjecting it to the most cursory of sanity checks.

In the current case, there is nothing that seems completely impossible, but there are some aspects which mean we're needing to use the plastic equivalent of unobtainium or wishalloy – wishalkene, if you will. But we can make some steps towards identifying the sorts of properties that we need for this to happen and we can also identify some real-world examples that will explain how this could occur, either through a series of unlikely mutations/genetic engineering or, more likely, a highly adaptive exo-suit. In terms of the properties, flexibility is obviously crucial, as is being able to withstand shock-loading, along with being adaptable/tune-able and possessing stiffness. So, how can this be achieved? Well, the work on memory materials (in the context of ideal ropes) looks very interesting, and there is some exciting work being done on programmable polymers which can change shape if left for sufficient time – they can even be programmed to move through a sequence of shapes, and to spend a given time in each form.[9] The problem with these programmable materials is that they mooove veeery sloooowly. However, there has also been some interesting work using a combination of superhydrophobic and superhydrophilic polymers to mimic the effect of *Mimosa pudica*;[10] these materials have been able to reshape themselves following external stimuli within ~30 milliseconds. And we have the real-world example of muscular hydrostats to draw on. On its own this is unlikely to be the answer, but as part of a tool-kit to design a suitable musculature, this is definitely something that we want to keep in mind.

Recalling what we said about plastic deformation, shape-memory materials offer a way to recover original properties after plastic deformation occurs. This might offer a way of producing a stiffened synthetic muscular hydrostat which could react quickly but still have enough stiffness and strength to provide superstrength (for when you pick your coffee cup up from 30 m away with an elastic arm) and resilience to (severe) shock loads. Exploiting this for extreme situations is still likely to be challenging though, especially for heroes (super or otherwise) for whom extreme situations are the norm (which is why the engineers that support them always seem so harassed and on the verge of developing an ulcer).

Finally, an answer to a question that cropped up during the course of the weekend whilst writing the book. How easy was it for Elastigirl to give birth? Based on what we have discovered about her physical capacities, it was probably very easy, as she is much more flexible than the average human. We've also established that she must have some sort of skeletal structure, including a pelvis. However, there's also a lot of 'mechanism' behind birthing, and midwives are very focused on the position of the baby, because the baby has to do some acrobatics of its own in this process. Helen Parr does have relatively broad hips (there is one moment in the film when she looks at herself in the mirror rather ruefully in respect of this) and that would make things easier for her. On the other hand, one of the preparations that the human body naturally makes for childbirth is the release of a hormone called relaxin. The role of relaxin (which is produced by the placenta) is to mediate some of the effects of pregnancy on the body, and to increase the flexibility of joints by causing ligaments to relax. It also prepares the uterus for the action of oxytocin. In some causes, the relaxin is too effective and the pelvic girdle becomes too loose, a condition called symphysis pubis dysfunction (SPD). SPD can be incredibly painful, with a certain amount of bone grating on bone and so on. It usually disappears with the other effects of pregnancy as the hormone clears from the body, although this process can take several months. So, on balance, it is likely that Elastigirl experienced a relatively quick, pain-free labour, but not without complications in the last few months of pregnancy.

So what superpower would you go for? In this chapter, we've considered some of the physical possibilities of superpowers and examined the likelihood of them being effective. We've also looked carefully at the trade-offs that superpowers often bring – having any given power may be great for a set number of tasks, but not necessarily optimal for everything we'd like to do. In this respect, reversibility might be one of the key aspects of any power, but this may be too much to ask.

REFERENCES

1. D. Ronca, *What if you really ate 50 eggs like in 'Cool Hand Luke'?*, HowStuffWorks, 2016, [cited 13 November 2016], available from http://science.howstuffworks.com/science-vs-myth/what-if/what-if-ate-50-eggs-like-cool-hand-luke.htm.
2. *Octopus Escape*, Video.nationalgeographic.com, 2016 [cited 13 November 2016], available from: http://video.nationalgeo-graphic.com/video/octopus_cyanea_locomotion.
3. K. C. Nishikawa, W. M. Kier and K. K. Smith, Morphology and mechanics of tongue movement in the African pig-nosed frog *Hemisus marmoratum*: A muscular hydrostatic model, *J. Exp. Biol.*, 1999, **202**, 771–780.
4. H. Parr, *Disney Wiki*, 2016 [cited 13 November 2016], available from: http://disney.wikia.com/wiki/Helen_Parr.
5. H. Petroski, *Invention by Design*, Harvard University Press, Cambridge, Mass, 1st edn, 1996.

6. D. Harutyunyan, G. W. Milton, T. J. Dick and J. Boyer, On ideal dynamic climbing ropes, *Proc. Inst. Mech. Eng., Part A*, 2016, DOI: 10.1177\1754337116653539.

7. M. Ashby and D. Jones, *Engineering Materials*, Elsevier Butterworth-Heinemann, Oxford [UK], 1st edn, 2005.

8. J. Gere, *Mechanics of Materials*, Brooks/Cole, Pacific Grove, CA, USA, 1st edn, 2001.

9. X. Hu, J. Zhou, M. Vatankhah-Varnosfaderani, W. Daniel, Q. Li, A. Zhushma, A. Dobrynin and S. Sheiko, Programming temporal shapeshifting, *Nat. Commun.*, 2016, 7, 12919.

10. W. Wong, M. Li, D. Nisbet, V. Craig, Z. Wang and A. Tricoli, Mimosa Origami: A nanostructure-enabled directional self-organization regime of materials, *Sci. Adv.*, 2016, 2(6), e1600417.

CHAPTER 12

Big Data, Big Heroes and Bad Computers

ROB MILES

University of Hull, School of Engineering and Computer Science,
Hull, HU6 7RX, UK
E-mail: R.S.Miles@hull.ac.uk

12.1 INTRODUCTION

Batman is definitely an early adopter. Well before the PC was invented in the 1960s, Batman's cave housed his very own Bat-computer[†] complete with Bat-input slot (scanner and character recognition software), criminal Bat-locator (GPS tracker) and Bat-slides (screen). Since then, computers have been a big part of superhero storylines and data-processing devices have served good and evil in about equal measure. The Bat-computer and

---

[†]Sharp eyed fans of 1960s and 1970s sci-fi may have noticed the 'Bat-computer' cropping up in fairly regularly on TV and movies from the time. It was actually a Burrough Corporation B205 computer from the 1950s, which by the mid-1960s was obsolete. However, all its blinking lights and whizzing reels of tapes made for a great prop and so it became a feature in *Batman, Lost in Space, The Fantastic Voyage* and more. It even made a come back in the late 1990s during *Austin Powers: The Spy Who Shagged Me*.

The Secret Science of Superheroes
Edited by Mark Lorch and Andy Miah
© The Royal Society of Chemistry, 2017
Published by the Royal Society of Chemistry, www.rsc.org

its add-ons seem to have been pretty good predictors of some of the technology that we take for granted now, but how close to real life are the descriptions of computer use in superhero stories generally? Can we look back on these stories and discover insights into what have become some of the biggest challenges and opportunities arising from the rise of supercomputers?

12.2   WILL ARTIFICIAL INTELLIGENCE ENSLAVE US ALL?

Mary Shelly's *Frankenstein* (published 1818) may be the first fictional account of the human race trying to create artificial life, and the story doesn't end well. Nor does the next account of artificial intelligence, 100 years later in a 1921 play by Karel Čapek called R.U.R (*Rossumovi Univerzální Roboti*[‡]). The play tells of a factory that manufactures artificial people called Roboti. Following a theme that is now familiar to any sci-fi fans (from X-Men's sentinels to *Terminator's* Skynet), the robots rise up against the human race. Since then, the human race's interactions with AI have been explored in great depth. Taking Mary Shelly's lead, it often doesn't end well for us.[§]

Superhero movies, as much as science fiction writing, often portray artificial intelligence as a threat to humanity. A great example of this is *Avengers: Age of Ultron* (2015) where – spoiler alert – a program designed to promote world peace is subverted into something much nastier, with the expected special effects-laden results. So, how likely is it that rampant software will try to take over from us humans? And what makes good software go bad?

12.3   COMPUTERS AND PROGRAMS

Before we can consider what makes a rogue program, we need to take a look at what defines a program. You can regard a computer as something a bit like a sausage machine. Stuff goes in one end and something else comes out of the other. In the case of a sausage machine, the processing that takes place converts meat into sausages. In the case of a computer, the stuff going in is data (perhaps a request for a bank loan), and the stuff coming

[‡]It was R.U.R. that introduced the word 'robot' to the sci-fi world, but the word is derived from Czech 'robota', meaning forced labour.
[§]As shown by a quick perusal of the Wikipedia page on artificial intelligence in fiction.

out is more data (perhaps a message responding to the request). The processing that takes place in a computer is performed by a program. A program itself is a sequence of instructions along the lines of 'If someone called Rob Miles tries to borrow a million pounds from this bank, then just say "no"'.

Programs are created by people with the surprising name of 'programmers', and it is their job to figure out what the program needs to do and then write a sequence of instructions that perform that task. Programs are also called 'software' because they are supposed to be easier to modify than the physical 'hardware' of the computer that runs them (although I've not always found this to be the case, if I'm honest).

One of the fundamental principles of programming is that you cannot write a program to solve a problem if you don't know how to solve the problem yourself. In other words, if you don't know how to take over the world, you can't write a program that will take over the world for you. This fact would seem to make the world safe from evil software, but it is not quite as simple as this because we have a branch of Computer Science called 'artificial intelligence'.

12.4   ARTIFICIAL INTELLIGENCE – THE RISE OF THE 'COMPUTER BRAIN'

There are lots of definitions of artificial intelligence, but let's keep things simple. I think that artificial intelligence is anything shown by a program that makes it intelligently useful. Anytime you find yourself being chased by a missile when playing a video game, you are seeing artificial intelligence in action. The program controlling the missile knows where you are and makes the missile accelerate in your direction. This is a reasonable definition of artificial intelligence, because the software is behaving in exactly the same way as you would if you were trying to chase me. In this sense, its alignment with human intelligence is often used as a benchmark for determining its status, which is a test that not all scientists think is ideal.

In the case of the missile in a video game, the intelligence knows the target's location, so the behaviour is very easy to create. However, to create a useful artificially intelligent device the computer might have to work a bit harder to make sense of the data that it is receiving. One situation where our software has to work with a lot of incoming data is when we try to create an artificially intelligent system that can make sense of human speech.

12.5   SPEECH RECOGNITION AS AN APPLICATION OF ARTIFICIAL INTELLIGENCE

Lots of superhero computers can understand what we say. Even the 'Bat-computer' of 1964 was voice controlled. In fact, this is one situation where we have caught up with the superheroes, as we routinely talk to our devices today. However, understanding language is hard to do with the 'sausage machine' model of a computer program. There's just too much data to process. So, to solve this and other complex problems, software designers have taken to creating systems that work by replicating the way that the human brain processes data.

12.6   HOW YOUR BRAIN WORKS

In your brain a huge number of interconnected 'neurons' are continuously processing data coming in from your senses. Each neuron has many inputs, some from the outside world, some from other neurons. Some neuron outputs are connected to nerves that control muscles so that our brains can actually make us do things.

You can think of each neuron as a tiny processor that looks at the values of its input signals and decides whether or not to 'fire' and produce an output signal. As waves of input signals pass through the massive network of neurons that is your brain the signals are processed into the sensible and rational behaviours that are characteristic of all us humans. Or perhaps that is just me ...

I must admit that this description is a massive simplification of what actually goes on and, in fact, nobody really knows what happens when we think about something and then perform an action. Something in my brain is causing me to press the keys on the keyboard and write this text, but I'm not really sure what happens in my brain to ensure this takes place. And to be honest, it hurts if I think about it too hard. However, we do know enough about the neural nature of the brain to be able to try to mimic this behaviour and make our own electronic version.

12.7   BUILDING ELECTRONIC BRAINS

We build 'brain-like' software by creating a program that builds a network of artificial neurons and updates each one in turn. When the program runs, the idea is that it will produce the output that we want in response to a particular set of data. In other words, if the program is given a sound sample of me saying 'cheese' it will produce the output 'cheese'.

A crucial component of the way the brain works is the behaviour of each of the individual neurons; what set of inputs causes a particular neuron to 'fire' and trigger other neurons. The neurons have to be 'trained' to pass on signals at the right time. Humans perform this training during a process known as 'growing up'. We are all born with a set of 'hard-wired' behaviours and then spend our childhood forming new ones based on experience.

We have to do just the same with our artificial brains. Speech-recognizing artificial intelligence is created by 'training' a neural network with a series of known sound patterns and giving mappings to what they mean. During training, the program adjusts the behaviour of the individual neurons in the network to produce the most accurate recognition of all the sample inputs. Then the program is given speech input and will make sense of it for us.

12.8   THE NEED FOR SPEED

The more neurons you can hook together in your electronic brain, the better the chance of the system doing what you want. The power of a real brain lies in its huge number of neurons and the way that they all operate at the same time, whereas until fairly recently a computer could only contain a few processing elements.

However, that has changed with the arrival of hardware which contains massive numbers of connected processors. You find these configurations in modern video-gaming hardware among other places. The latest games consoles contain over a thousand individual processors. They are used to spread the load of creating the impressive graphical environments of modern games.

By using very large numbers of such processors to create large 'deep neural nets' with lots of neurons we can greatly improve the performance of our artificial intelligence-powered solutions and create genuinely useful intelligent devices.[1]

12.9 BROKEN THOUGHTS

This all sounds rather wonderful. We can build systems that will work the way that our brains do, and create properly useful intelligent devices. But it turns out that this form of artificial intelligence has one serious limitation – it is only as good as the training process. There's a story, probably apocryphal, that illustrates what can go wrong.

Apparently, towards the end of the last century the US military got very excited about neural networks and decided to build a system that could spot military vehicles that were hidden in forests. They created a neural network and fed it 200 pictures of forests with tanks hidden in them, and 200 with no tanks. The network was then laboriously trained to identify the pictures that contained tanks. In other words, the behaviours of each neuron in the network were adjusted so that the program gave the correct output from each picture. Eventually the network was trained up to the point where it seemed to be working perfectly.

If you showed the network one of the pictures with tanks in, then it said 'Tanks'. If you showed it one of the pictures with no tanks, then the program said 'No tanks'. All good. But, then they tried some new pictures that the system hadn't seen before, and its performance was rubbish. It just didn't work.

After a lot of head scratching they discovered that all the pictures with tanks in them had been taken on a cloudy day, whereas all the tank-free pictures had been taken on a sunny day. What the military had actually created was a program that could identify pictures taken on sunny days. This illustrates the problem with artificial intelligence. At the end of the day, a system is only as good as the training that it has received. If what you give it doesn't match any of the training data, then you might not get what you expect. To make matters worse, when a neural net has been trained up, nobody really understands the process by which it reaches its conclusions. This means that it is not something you can fix if it breaks, since you don't actually know how it works.

To bring this right up to date, the photo sharing service Flickr® can automatically categorize pictures using technology that works in this way. It's very good, but it is not perfect. Look at Figure 12.1.

This is not a picture of food. But Flickr® thinks it is. Some elements of the picture do conform to what some foods look like,

Figure 12.1   What is this? Image credit: Rob Miles.

but even a very young child would know that there was nothing in the picture you could eat.

So, it is unlikely that present-day artificially intelligent systems will decide to destroy or enslave humanity, but they should be treated with a serious level of caution as their 'guessing'-based behaviour might lead them to do the wrong thing every now and then. This is a particular problem when we use this kind of intelligence in safety critical systems, for example flying a plane. The auto-pilot may have been trained to fly under all *normal* conditions, but might not react sensibly when faced with something it hasn't seen before. In that case we need to be able to hand control back to an experienced pilot who can use their depth of human experience to deal with the situation. There have actually been crashes which have been caused by the auto-pilot 'giving up' and handing control back to a pilot that wasn't prepared for this to happen.

So, just like lots of other tools that we have created over the years to improve our lives, artificial intelligence is useful, but we have to be aware of its inherent limitations.

12.10   BUILDING UNDERSTANDING INTO MACHINES – HOW DO WE MAKE 'JARVIS'?

In the Iron Man series of films Tony Stark has 'Jarvis' (Just A Rather Very Intelligent System) as his home computing system. Jarvis is able to understand everything that Tony says, and can even engage in banter and come up with original thoughts. When will we get computers that can engage in conversation like this in real-life?

It turns out that, while voice recognition is now eminently doable and generating high-quality speech is also within the reach of even quite modest computers, making a computer understand what we are saying and respond in a 'sensible' way is still a very difficult thing to do. Companies such as Amazon® and Google® have recently launched such tools in the form of their 'Alexa®' and 'Home' assistant robots.

12.11   HOW DO WE MAKE A HUMAN COMPUTER?

One of the fathers of modern computing, Alan Turing, proposed a test for intelligence in the 1940s which became known as the Imitation Game. His test was both pragmatic and simple, and involves three parties – a man, a woman and an interrogator. The interrogator is not able to see the man or the woman, but through text-based conversation must determine which is which. In the first part of the game, the interrogator plays with people, but in the second part, either the man or the woman is played by a computer, which is trying to appear as either a man or a woman. The hypothesis is that, if the interrogator has no variance of success between part 1 and part 2 of the game, then we can conclude that the computer is intelligent.

Turing's theory spawned the annual Loebner Prize, where judges talk to computers and people and try to tell them apart. It turns out that a really good way to 'cheat' when writing a 'human' program is for the software to adopt a particular persona for the conversation. If a program can draw the human protagonist into its world, then it can provide a reasonable facsimile of that world. One very early program, Parry, behaved like a paranoid schizophrenic and did this quite convincingly. Loebner prize winners have adopted a variety of different personalities over the years.

However, these systems are still easily fooled. If you ever find yourself wondering if you are talking to a computer or a person, your best bet is to say something really stupid and see what happens. Enter something like 'I've just eaten a house for breakfast' and see what comes back. If the response doesn't express some degree of surprise at what you just did, then you might start to become suspicious that it may be a machine on the other end of the line.

12.12 BUILDING UNDERSTANDING

Of course, Jarvis could pass the Turing Test without even breaking a sweat, but he also has the ability to understand what Tony Stark wants when they have a conversation. Passing the Turing Test just involves constructing a response that makes sense to the other party in the conversation – it doesn't mean that the computer program actually has to understand what you want.

However, the good news is that computer systems are now getting powerful enough to both engage in rudimentary conversation and understand want we want and act on it. Personal assistants such as Apple®'s Siri®, Microsoft®'s Cortana®, the Google® Now® service and Amazon®'s Alexa® are becoming genuinely useful as they harness powerful artificial intelligence techniques to extract meaning and context from the questions that we ask. You might not be able to engage in conversation about the meaning of life with these systems (although they do tend to reply to this sort of question with a reasonable quote from *The Hitch Hiker's Guide to the Galaxy*), but they will be able to give you the time of the next bus home, or whether you need to take an umbrella when you go out.

In fact, the chances of us each having our own personal 'Jarvis' in the future are actually quite high, but we need to remember that, although these devices are extremely useful, they don't just serve us – they may also have a much higher calling – something we'll refer to later when we consider the 'software bubbles' that engulf our lives.

12.13 WHEN GOOD SOFTWARE GOES BAD

In a lot of science fiction and superhero storylines we find that the computer is not a 'bad' guy as such because, unlike us, it doesn't really know the difference between wrong and right. Take one of the first AI's to go bad on the big screen, the notorious HAL 9000 in Kubrick's epic *2001: A Space Odyssey* (which incidentally gets an A+ for accurate science in a sci-fi movie). Homicidal HAL, whilst in charge of the spacecraft *Discovery One*, tries to kill all the crew on board when he decides that he could do a better job of running the mission than the humans can. But, maybe he has been badly mis-represented. Poor old HAL never received any ethical programming. So, it's little wonder he starts exterminating the crew when he realizes he can do a better job of running the mission than the meat bags he's carrying around can. A moral person may not have come to that conclusion, but a program would only behave ethically if this behaviour has been specifically added. Put simply, it's the programmers' fault that HAL refused to open the pod bay doors.

12.14   GIVING SOFTWARE AN AGENDA

There's a lesson to be learnt from HAL. We might encounter serious issues with artificially intelligent systems when we give them aims. Conceivably, we could create an artificially intelligent machine and tell it to maximize the stock price of a particular company. We could use an improved design for our neural net, creating software that 'learns' by changing the settings on the neurons based on the experiences that it has. We could hook the AI up to internet-enabled devices and systems and then we could end up in a bit of a moral panic, as there would be no guarantee that all the behaviours that emerge will be ethical. Our system might shut down factories, throw people out of work and maybe even start wars to achieve the set aims. As soon as you start to build a machine that affects the world around it, you have to start thinking about questions of ethics.

This turns out not to be a new idea. Way back in the 1950s a bio-chemist-turned-science fiction author sat down with his editor and discussed how an AI could be made to behave in an ethical way. At the time everyone viewed artificially intelligent devices as 'robots', and so what emerged was Isaac Asimov's Three Laws of Robotics:

1. A robot may not injure a human being or, through inaction, allow a human being to come to harm.
2. A robot must obey the orders given it by human beings except where such orders would conflict with the First Law.
3. A robot must protect its own existence as long as such protection does not conflict with the First or Second Laws.

These laws make a lot of sense and have underpinned a large number of successful novels and short stories, not to mention one or two films. The laws are also quite useful today, as a starting point in developing an 'ethical robot'. However, it turns out to be quite easy to take an ethical robot and change it so that it behaves in an unethical way. All one needs to do is change the perceptions of the robot to turn a good behaviour into a dangerous one. Remember that, as far as a computer program is concerned, it regards its entire universe in terms of the inputs coming in and the outputs that it produces. If the program 'thinks' it is controlling toy guns but, in fact, it is shooting at real people, then it would be frighteningly easy to assign it evil tasks. The film *WarGames* (1983) explores a variant of this possibility where the WOPR (War Operation Plan Response) computer decides to instigate nuclear war as opposed to a nice game of chess.

Yet, there are good reasons to doubt the possibility that artificially intelligent beings will ever take over the world on their own, but it could be a very useful tool for the bad guys to use. I don't lie awake at night worrying about a rogue program deciding to enslave humanity (why would it?), but I do worry about powerful computer systems and big data sets in the hands of people with less than benevolent aims, so let's look at these next.

12.15 HOW CAN YOU USE COMPUTERS TO TAKE OVER THE WORLD?

The plot of the 1983 *Superman 3* movie had an industrial tycoon harnessing software for world domination. How plausible is this? What do modern computer systems provide for the aspiring world domineer?

12.16 BIG DATA AND READING THE MIND OF HUMANITY

In the X-Men movies, Professor X has the power to read and control the minds of others. While this is not quite possible today, there is technology in development and application that performs a role that gets close. For instance, work by Kevin Warwick has explored brain-to-brain communications with some success, and researchers at Washington University were able to send brain signals from one person to another.[5] They do this by monitoring the brain for specific kinds of activity and, when noticed, trigger a signal to the other brain *via* transcranial magnetic stimulation. If this seems a step too far, then it might be possible to achieve similar ends by using big data to track what people do and 'read the mind of humanity'.

The world is generating data at an unprecedented rate. In 2016, *YouTube*® users uploaded around 400 hours of video every minute, and each hour of video required around 1 Gbyte of storage. In other words, there is enough video being uploaded to fill the average smartphone in five or six seconds. And the rate at which video is being uploaded is increasing by a factor of 10 every five years.

Of course, video is just one form of data. There is also the huge amount of data generated by the devices that we use, as they transmit our location and activity. Companies such as Facebook® and Google® don't just throw away these data, instead they are spending millions of dollars each year on hard disk storage and the physical design of hard disks is being changed to reflect the needs of their mass data storage.[2]

All this 'big data' can give incredible insights into the minds and behaviours of the populations who are generating it. Mostly we don't seem to mind because the results can be pretty useful. How do you think Google® Maps® is so good at telling you where traffic jams are? Simply because Google® is checking for where there's a load of stationary mobile phones in the middle of roads. Big data has the rather unhelpful definition of 'Any data set which is too large to be processed using conventional data processing technology'. Big data is a moving target, as both the size of datasets around the world is growing, along with the power and capacity of the systems that can process it. However, by employing large numbers of connected computers it is possible to draw useful information from even the largest sets of data.

A technique called 'data mining' can be used to profile users and extrapolate their interests from their profile. You see this in music players, when platforms decide that, if you listen to the Beatles rather a lot, then you might like to hear some tracks by the Beach Boys. This is based on the data insights that show how other Beatles lovers have also clicked on a large number of Beach Boys tracks. Or, if you suddenly start searching for baby buggies, then the systems may decide that you would also like to know about playpens and other baby-related items.

This is actually happening now, in that the view of the world that you see *via* your computer is different from that of anyone else. The results of your searches and the items suggested for you to look at are all coloured by your previous actions. If I search for 'Rob Miles' I see my own web pages at the top of the list of results, but this is not the case for everyone. Another 'Rob Miles' will probably see their name at the top.

This is actually quite a dangerous development as it locks us into a 'filter bubble', which means we only ever see the things we want to see. This might matter from an educational perspective, as it is often confrontation with something we don't believe that pushes us to think more critically. Instead, the filter bubble leads us to believe that our views are always right, because we've never found anyone who thinks differently from them.

Furthermore, your friendly helpful personal assistant also learns from your searches and may also be made to deliver some results in preference to others, depending on the agenda of the system hosting that particular service. I can't find any superhero movies that use this as the basis of a plot against humanity, but you may know more than I in this respect.

12.17 TURNING OUR DEVICES AGAINST US

We are surrounding ourselves with more and more devices which contain computers. Can these be made to turn against us? The best recent movie example that I've found of this happening is the movie *G-Force* (2009),[3] in which a team of superhero guinea pigs (really) thwart an attempt to destroy the world. The attempt planned to cause household appliances to mutate into killer robots. Is it likely that our toasters will turn against us, and what can we do to prevent it?

12.18 THE INTERNET OF BAD THINGS

It is scary, but true, that this kind of thing has already happened. In October 2016 very large parts of the internet suddenly became inaccessible. Services like Netflix®, Spotify™, Twitter® and PayPal® failed for millions of people in the United States and around the world. Attackers broke part of the internet.

To understand how this happened you have to remember that the internet was designed a long time ago, when there were many fewer computers in use. The original aim of the internet was to build a network that would survive nuclear attack. It was important to the customers of the original internet (the US Military) that if parts of the network were damaged or destroyed all the messages would still get through. The internet breaks messages into small packets, each of which is individually routed from one system to another. If one part of the network is vaporized by an atomic blast, then the systems automatically re-route data packets *via* the remaining parts.

You might think that this would make the internet more resilient to attack, but unfortunately this is not the only way that you can attack a network. The original designs for the internet reflected a great amount of trust by the designers in the good behaviour of the connected systems. Since, in the original network, all the systems were owned by members of the military or their friends, there was not a lot of attention paid to the possibility that the computers themselves might do bad things.

The kind of incidents that we saw in October 2016 revolve around the mechanisms that the internet uses to resolve the physical addresses of systems on the network. Each system on the internet has a unique address, a telephone number if you like. But users of the internet like to refer to a destination by its name, rather than a number. We don't want to have to remember 192.168.15.23. We want to use www.netflix.com. The process of converting a name to its corresponding network address is performed by a component of the internet called the Domain Name System or DNS.

When you type www.netflix.com into your browser your computer askes its local DNS server to give it the network address corresponding to that name. If the local server has a copy of the address it will pass it straight back. If it doesn't know the address it asks another DNS server for the address. If that server doesn't know the name it asks another server, and so on until a system is found that does know the answer. The system is cunningly designed so that addresses which end in .com are managed by a different server from those that end in co.uk, so that different sets of addresses can be managed by different servers.

DNS management can be tricky, particularly for very popular web sites, so they frequently employ another company to do this for them. One popular company is Dyn®, and it found itself the victim of a 'Distributed Denial of Service' attack in October 2016.

A 'Distributed Denial of Service' attack on a service works by bombarding the target with a huge number of requests at the same time. The idea is to overwhelm the service and stop legitimate traffic getting through. It's rather like every person in the country calling Directory Enquiries at once. If we did that the service would become unusable.

The designers of the internet never thought that large numbers of rogue systems would be used in this way. Furthermore, at the time the internet was designed a computer powerful enough to be connected to the network cost many thousands of pounds, so getting enough hardware to mount an attack on the network 'from inside' would have been a very expensive proposition. However, fast forward to today and we have a situation where anyone can connect a system to the internet and start sending messages.

In fact, things are even worse. A great many things connected to the internet are extremely insecure and open to malicious use. Devices such as internet-connected security cameras are actually complete computer systems. Unfortunately, they are supplied with fixed usernames and passwords that their owners don't often get around to changing. So it's easy for the bad guys to take control of these devices and then use them to attack whatever target they fancy. Also, it turns out that a software package that can perform the infection and then manage the resulting network of remote-controlled 'bots' is easy to find for anyone to use.

In fact, some people[4] have postulated that the some of the recent activity looks like it is state sponsored, and some countries are trying to add the capability to take down the internet to their arsenal of weaponry – a deeply troubling development.

This means that we can expect future wars to be fought on the internet, in the same way that the Tony Stark's Jarvis was attacked in the movie *Avengers: Age of Ultron*. Fighting such a war is very difficult. The attacking software will make it hard for services to detect the rogue messages and block them. We just have to wait until the design of the internet is changed to deal with this problem and people learn that many of the gadgets that they are buying are actually fully fledged computers that need a bit of system management.

This is a situation where superhero stories are starting to come true, and not in a good way. It's very unlikely that the webcam in your study will suddenly grow legs and chase you around the house. However, unless you take proper precautions, your home devices could be used in an attack on the backbone of the internet. So, you can play your part in a superhero battle just by making sure that all the intelligent devices that you own have good strong passwords.

REFERENCES

1. A. Graves, A. Mohamed and G. Hinton, *Speech Recognition with Deep Recurrent Neural Networks*, Arxiv.org, 2017 [cited 8 November 2016], available from: https://arxiv.org/abs/1303.5778.
2. E. Brewer, L. Ying, L. Greenfield, R. Cypher and T. T'so, *Disks for Data Centers*, Research.google.com, 2017 [cited 8 November 2016], available from: https://research.google.com/pubs/pub44830.html.
3. http://www.imdb.com/title/tt0436339/, 2017 [cited 8 February 2017].
4. B. Schneier, *Someone Is Learning How to Take Down the Internet–Schneier on Security*, Schneier.com, 2017 [cited 8 November 2016], available from: https://www.schneier.com/blog/archives/2016/09/someone_is_lear.html.
5. R. Rao, A. Stocco, M. Bryan, D. Sarma, T. Youngquist and J. Wu, *et al.*, A direct brain-to-brain interface in humans, *PLoS One*, 2014, 9(11), e111332.

The Wonder-ous Truth: The Workings of Wonder Woman's Lasso

FELICITY HEATHCOTE-MÁRCZ

Alliance Manchester Business School, University of Manchester, Booth Street East, Manchester, M13 9SS, UK
E-mail: flicstyle@hotmail.com

13.1 WHAT IS THE LASSO OF TRUTH?

While we often focus on the capacities of superheroes, there are also extraordinary abilities that are made possible by the gadgets they have. This chapter gives careful consideration to fully understanding how Wonder Woman's lasso works. With names such as the Golden Lasso, the Magic Lasso, or the Lasso of Truth, this mythical instrument was created by William Moulton Marston, when he wrote the first Wonder Woman comics in the early 1940s. The Lasso of Truth has several origin stories: it was a gift to Wonder Woman from her mother Hippolyta; it was forged from the God Hephaestus from the Golden Girdle worn by Hippolyta's sister; it was made and took its powers from the fires of Hestia,

The Secret Science of Superheroes
Edited by Mark Lorch and Andy Miah
© The Royal Society of Chemistry, 2017
Published by the Royal Society of Chemistry, www.rsc.org

the most ancient God of Olympus; it was given to Princess Diana before she left Paradise Island to live as Wonder Woman, or after she left. However, these stories share the fact that they present the Lasso of Truth as an incredibly powerful piece of mythical equipment bestowed by the gods of Ancient Greece, and made for supporting the causes of truth and justice *via* its magical properties.

Anyone who may become caught in the Lasso of Truth, or who even touches the Lasso in some cases (The Green Lantern making a faux pas for instance),[1] will be compelled to speak truthfully and will be unable to hide their deepest motivations. This Magic Lasso holds other powers too, such as the capacity to cure insanity (such as for Ares when he tries to start World War III[2]), and to quash even the power of atom bombs (Wonder Woman Volume 1 #14). The Magic Lasso can also expose illusions,[3] can kill DC Comic's un-dead incarnations the 'Black Lanterns',[4] and can render those caught in the Lasso's grip under ultimate mind control (Superman and Brainiac being the most famous examples). The second season of the Wonder Woman action series also sees the Lasso acquire the power to cause selective amnesia for its captives, and in the Super Friends animated TV series, Wonder Woman is seen communicating telepathically with her golden weapon.

The rope is said to be golden, unbreakable and of infinite length.[5] However, there are occasions in the DC Comic's world when it has been broken and I'll discuss these extraordinary events in more detail later in this chapter. For now, let's turn to the superhero who has wielded the Lasso of Truth over many decades and hundreds of comic book issues: Wonder Woman.

13.2 THE WOMAN BEHIND THE ROPE

'Beautiful as Aphrodite, wise as Athena, swifter than Hermes, and stronger than Hercules.' Wonder Woman's strapline description which appeared at the start of every issue of the DC comics' superheroine, says much about this powerful character. Yet, these descriptions also leave us with many more questions and intrigue. With a set of supernatural abilities and an array of novel gadgets for defeating evil at her disposal – and being the first female hero to disrupt the male-dominated comic-book landscape of the 1940s and arguably still today – Wonder Woman commands a mystique and interest unique to the superhero world. Her origins are ultimately unknown. We don't know how she became so powerful. And, most importantly for this chapter, we don't really understand the magic Lasso, which she uses to gain the truth from her captives. These unknowns have mystified fans and critics alike, since our mysterious and wondrous (or wonder-ous to emphasize the etymology as 'full of' wonder) woman first appeared in her 1941 comic debut.

Wonder Woman is a superhero of multiple identities, alter-egos and disguises, as was the case for other DC Comics characters during its 'Golden Age' (c. 1938–1947), namely Batman and Superman. Originally, Wonder Woman was a princess of the Amazonian people, a mythical tribe of warrior women conceived in Roman times. Wonder Woman was known in her homeland as Diana of Themyscira. This link to the mythic Goddess Diana, goddess of nature, hunting and women, describes well Wonder Woman's enduring reputation as the 'ultimate woman' and feminist icon she has come to embody. For Wonder Woman's creator William Moulton Marston, Themyscira (or Paradise Island) was a home of love, safety and autonomy for the Amazon female warriors, and represented freedom from male bondage and the patriarchal rules of 'Man's world' (Earth).

Outside of this paradise and when concealing her super-abilities, Wonder Woman is known as Diana Prince and she takes on many roles, such as a government spy, nurse and secretary, and even a boutique store owner when she loses all her superpowers in the 1960s and 1970s. As such, this multi-faceted woman is something of an enigma, and also a political football for her various writers throughout the decades. From a symbol of the suffragist movement of the early 20th century (William Moulton Marston was an undergraduate at Harvard when famous suffragette Emmeline Pankhurst was banned from giving a speech there), to a 'girl next door' figure portrayed by writer Gail Simone in the 1970s as 'a very relatable and sympathetic character',[6] to Gloria Steinem's second wave heroine gracing the cover of *Ms* magazine with full restoration of superpowers, Wonder Woman has changed with the times and reflected the cultural milieu of every age.

In more recent years, Wonder Woman has taken on a harder edge as a Superhero prepared to use deadly force (in her killing of Maxwell Lord[7]), as a bisexual trailblazer for DC comics, according to current Wonder Woman writer Greg Rutka,[8] and as the controversial UN 'honorary ambassador for the Empowerment of women and girls'.[9]

More than any other of DC Comics' classic characters, Wonder Woman is a symbolic figure steeped in many layers of cultural references and ideological viewpoints. The meaning of her existence, origin and superpowers and the unique equipment she wears on her body (her body being another powerful and symbolic weapon on many levels) to defend herself and fight her enemies, brings with it competing claims over what she represents.

The mythological character of comic strips finds himself in this singular situation: he must be an archetype, the totality of certain collective aspirations, and therefore, he must necessarily become immobilized in an emblematic and fixed nature which renders him easily recognizable. (Umberto Eco and Natalie Chilton, *The Myth of Superman*, Diacritics, 2, no. 1, 1972, 15)

This quote from literary critic Umberto Eco may describe well the fixed tropes and characterization of the most well-known male superheroes. However, it fails to recognize Wonder Woman's chameleonic nature; from her different outfits, abilities, love interests and sexualities, histories and ontological states (alive? dead? both?), she is fundamentally fluid and a figure rooted in the dynamics of change.

Her creator is claimed to have said: 'Wonder Woman is psychological propaganda for the new type of woman who should, I believe, rule the world.' What he meant by this was that his superheroine was a figure to represent all the best traits women were traditionally assumed to naturally possess, along with all the best of those naturally assigned to men, and that this gender-bending fantasy of an all-powerful woman should inspire girls in the real world:

Not even girls want to be girls so long as our feminine archetype lacks force, strength, and power. Not wanting to be girls, they don't want to be tender, submissive, peace-loving as good women are. Women's strong qualities have become despised because of their weakness. The obvious remedy is to create a feminine character with all the strength of Superman plus all the allure of a good and beautiful woman. (Marston, 1943 issue of *The American Scholar*)

From Marston's description above and his other writings on the untapped power of the female ('the only hope for civilization is the greater freedom, development and equality of women'[10]), we can understand that, for him, women were the political answer to a world filled with hate, bloodshed and war. Therefore, it follows that it could only be a woman who would have the capacity to bring justice and truth in her allegory, body and spirit; that the superhero to make captives submit and speak with the wondrous Lasso of Truth could only be Wonder Woman.

13.3   TRUTH AND THE LASSO

The Lasso of Truth is the perfect instrument of the 'loving submission' concept that so fascinated Wonder Woman's creator, William Moulton Marston. It is also an important device for understanding how this inventor (pseudo-scientist) and comic writer understood the nature of truth and reality.

Some argue that the Lasso of Truth is a device through which Marston channelled his latent erotic ideas, especially his focus on bondage that had its basis in the psychological (or perhaps more accurately, pseudo-psychological) research he carried out for much of his early career (more on this later). However, others have argued the Lasso derives from Marston's enthusiastic use and evolution of the polygraph or lie-detector test.

During his undergraduate years at Harvard University, Marston's mentor and collaborator was a colourful character called Hugo Münsterberg, a German emigrant to the US who was invited to set up a laboratory of psychology at Harvard and became well known to the American public for his outspoken views supporting the policies and politics of the German Empire during the First World War. Marston's eccentric tutor 'opposed votes for women and thought educating them was a waste of time'[11] and is widely believed to be the inspiration for 'Dr Psycho', a nemesis of Wonder Woman in Marston's comics.

During his time as Münsterberg's lab assistant, Marston is said to have invented the systolic blood pressure test, which later became an important component of the polygraph or lie detector, with the help of insights from his wife Elizabeth Holloway Marston. It is thought that Elizabeth Marston observed the changes in physiological states of research participants in their experiments as well as her own reactions: 'when she got mad or excited, her blood pressure seemed to climb',[12] which became a key insight in linking blood pressure changes to deception in William Marston's scientific books and articles. However, Elizabeth is not named as a collaborator on any of his works and this is perhaps an ironic and rather unfortunate fact for a man who claimed to be such a progressive advocate for women. This seeming hypocrisy is found also in the fact that he lived in a poly-amorous relationship with another woman, Olive Byrne, a situation which he is rumoured to have forced his wife to accept.

The work of these three on the systolic blood pressure test, as well as William Marston's experiments and theories of deception, directly inspired the concept of the Lasso of Truth in the Wonder Woman comics, which Marston would go on to create after his career in science failed to blossom. Indeed, Geoffrey Buhn sees the Lasso as a direct fictional manifestation of a polygraph test:

> If freedom was a product of discipline for Marston, then so was truth. Flexible as rope, 'but strong enough to hold Hercules!', Wonder Woman's 'Golden Lasso' was the most visible expression of this idea (Fleisher,1976: 210). Anyone caught in the lasso found it impossible to lie. And because Wonder Woman used it to extract confessions and compel obedience, the golden lasso was of course nothing less than a lie detector. (Geoffrey Buhn, 1997)

13.4 SUBMIT OR LASSO!

However, this description of the Lasso's origins may be too simplistic, as the idea of the Lasso of Truth may be more closely attributed to Marston's theoretical work on emotions and the themes of dominance and submission in the relationality between men and women. Marston's near-obsessive writings on the tension between dominance and submission had an obvious and explicit sexual undertone that did not go unnoticed by readers of Wonder Woman materials, and these themes remain pertinent in every Wonder Woman adaptation. However, this obsession was more than titillation and aesthetic fantasy. Rather, Marston saw the interplay between oppositional forces of the aggressive and the submissive as central to his understanding of psychology and, henceforth, to liberating the mind for a positive future. Marston's blood pressure experiments on men and women convinced him that women had greater capacity for emotion and understanding and he used the character of Wonder Woman as an experiment to communicate his allegorical ideal of womanhood to his readers – women should be dominant, yet kind, becoming what he saw as 'love leaders'.[13] Such benign, 'loving' figures of authority seemed to have captured Marston's imagination as the solution for social evils. Individuals should submit to such power willingly, or be taught by (loving, of course) force to do so. In this sense, being a superhero was not simply a matter of having a special kind of physical power, but perhaps also a moral power that exceeds those enjoyed by normal people.

Marston's conviction was that women held the power of 'love allure', which could overcome the aggressive masculinity of the current status quo; the unforgiving and hard authority that had led to the first and second world wars. This characteristic was antithetical to loving authority that Marston believed in so strongly. He understood the comic book world prior to Wonder Woman, which consisted of reflections on the violent lives of men, but rejected their 'blood-curdling masculinity'.[14] His revolutionary idea (again influenced by his wife Elizabeth) was to create a superhero who would conquer with love, rather than war, and who did not exhibit the masculine traits that had filled the adventures of previous comics. These ideas also led him to theorize that submission was a kind of love and gift to be cultivated, and his psychological experiments led him to believe that it was only women who could use their natural submissiveness to the advantage of the greater good:

'... the submissiveness of women does not mean that women must submit, rather, it means that men must learn from women how to submit lovingly'. (Berlatsky, p.116)

Taking this into account, it becomes clear that Marston transferred such ideas into his development of an extension of Wonder Woman: her Lasso of Truth. The golden rope binds those who she – the hero of women and justice – decides must submit and they are powerless to resist its clasp. This latter aspect may also be a metaphor for Marston's view that the seduction of feminine charm was also irresistible.

13.5   WHEN THE TRUTH IS NOT ADMISSIBLE IN COURT

The idea that the magic Lasso produces an infallible version of the truth might be a reason to doubt that its genesis was the systolic blood pressure or polygraph test. After all, when Marston attempted to use his test results as evidence in courtrooms, the public, the wider scientific community of which Marston attempted to become a part, and the judicial system were highly sceptical of its ability to tell the truth beyond reasonable doubt. In fact, Marston attempted to demonstrate that a man being convicted of murder in 1920 was innocent by carrying out a polygraph test on the suspect. However, the results were not admitted as evidence and Marsden's potential career as a lawyer was over. This must have been quite a devastating blow for him, and this traumatic experience of his invention being rejected by the justice system may be the more likely beginnings of the Lasso of Truth.

In one edition of the Wonder Woman comic,[15] we can see the Lasso of Truth being utilized by Wonder Woman to cross-examine a witness and extract 'the truth'. In this comic strip the judge calls the Lasso 'remarkable', and accepts it into the court despite the unorthodox character of the instrument. Perhaps Marsden yearned for this kind of acceptance when he presented his polygraph test results as evidence of a defendant's innocence, and that the literary device of the Lasso was a cathartic expression of Marsden's frustrated ambitions.

13.6   BREAKING THE LASSO

As previously discussed in this chapter, Wonder Woman's character changes with the historical seasons of her writers and the zeitgeist of her times. As such, it is interesting to contrast this relative 'truth' of the Lasso's master with the absolute 'truth' defined by the Lasso itself. In JLA #62, the Lasso was broken when Wonder Woman refused to accept a confession made by Rama Kahn of Jarhanpur. This suggests that the rejection of truth is enough to break the Lasso, and this leaves us with difficult questions as to the validity of truth in a world where there may be many versions of the truth (post-truth politics), and where truth may be socially constructed. It seems there is little room for such possibilities in the fictional and black-and-white world of the golden Lasso.

Another occasion when the Lasso's power has been shattered took place when the supervillain Bizarro becomes captured in the rope.[16] This character may be interpreted as a chastising of the surrealist movement of early 20th century Europe and its various 'bizarre' artists and figureheads. Bizzaro claims that there is no true reality and this belief is too much for the Lasso, causing it to snap. A similar scene occurs when the Queen of Fables becomes caught in the Lasso[17] and her ability to bring un-real characters to life also disables the power of the Lasso. This is particularly ironic in a comic full of fantastical characters and superhuman powers.

From this story, we learn of the Lasso's infallibility and how, for the writers of Wonder Woman comics at least, truth is real and is something which can be accessed and assessed by an instrument such as the Magic Lasso.

13.7   HOW MIGHT A REAL-LIFE LASSO OF TRUTH WORK?

Explaining how a golden rope like the Lasso might work outside of the fantastical world of Wonder Woman is no easy task! It would be easiest to say the fires of Hestia magically expel the truth from within the villainous captives of the Lasso, yet this may be difficult to justify empirically. However, there are some ways in which we can imagine the creation of such a gadget, based on real-world science.

13.8   NECTAR OF TRUTH

Our first entry point into explaining the Lasso scientifically draws on what we currently know to be possible in compelling humans to be more honest. In this field we have the possibilities of social coercion,[18] the polygraph test of which Wonder Woman's creator was such a fan, and the solution that is the most invasive but which is argued to be the most effective in expelling the truth: a drug that could act as a 'truth-serum'. For this option there are several drugs which could be chosen as the poison administered by the Lasso. Firstly, a high dose of alcohol could be used to induce drowsiness, impair orientation and a dull the brain's 'alarm signal' that the Lasso's captive was saying something they should really keep quiet about. However, alcohol is a fairly unreliable truth serum and research shows it does not affect our abilities to control our behaviour fundamentally.[19]

Alternatively, scopolamine is another drug for potentially extracting confessions from a Magic Lasso's captives. This drug, also known as hyoscine hydrobromide, is used to treat motion sickness, but it is known to have hallucinogenic qualities when taken in larger doses, and was used by totalitarian political regimes such as the Czech communist state in the 1950s and 1960s to force prisoners to confess their political crimes.[20] Scopolamine is also administered *via* a patch mechanism (*trans*-dermal application), which would be how a Magic Lasso would administer a drug to those trapped in its grasp. However, scopolamine has also been shown to induce many undesirable side-effects, such as fearful hallucinations, which may defeat the object of the Lasso if the truth was compromised by such effects.

However, the drug most commonly associated with provoking truth telling is thiopental serum, or sodium pentothal. This is the drug made famous by Hollywood spy thrillers and detective novels as the super-drug that can make those injected with it tell the truth. Sodium pentothal is a barbiturate general anaesthetic which has a rapid-onset effect and which lasts only a short time once administered.[21] This would be helpful for a Magic Lasso. This very strong drug is used for both euthanasia (Belgium) and lethal injection (United States), but when administered in small doses, it renders cortical brain functioning sluggish and less effective. The cortex regions of the human brain are involved in so called 'higher brain functions' such as memory, learning, perception and abstraction.[22] Other research has shown how people with demonstrably highly developed prefrontal cortexes are better at deceiving others.[23] Therefore, by affecting the ability of the cortical regions of the brain to function, sodium pentothal forces people to tell the truth, as they are unable to engage in the higher functioning brain activity of lying. However, similar to the polygraph test, randomized controlled trials have not demonstrated the effects of the drug to be sufficiently effective to be used as evidence in court. Yet, since this is the best drug we currently know of for coercively extracting the truth, this drug would be the best choice for the real-life magic driving the Lasso of Truth.

13.9   OUR 'PATCH-LASSO'

So now we have chosen our 'truth serum', how would this drug be administered to those caught inside the Lasso? Considering all the Lasso can do is touch the skin of its captives (it could not inject the drug into their veins or feed it to them, for instance), a 'Lasso of Truth' could be made based on the model of the nicotine patch: an outer impermeable layer and an absorbent layer of material would sandwich a drug that could force victims of the Lasso to become more truthful and make it more difficult for them to lie when asked questions. This is the model for how scopolamine is administered for motion sickness, hence we know it is possible for drugs with these properties to take effect *via* the skin. This 'patch-Lasso' may sound like a fairly simple mechanism, but the science behind such a Lasso would require some ingenious inventing to put each constituent part together and make the Lasso work (a job William Moulton Marston might have enjoyed!).

To begin with, we would need a super-strong material for the impermeable outer layer of the Lasso. It must also be resistant to water, chemical, flame and frost, and plaited in the pattern of a beautiful golden rope. Luckily, there is one material which fits the bill for such a varied task: aramid rope.

Aramid, otherwise known as high modulus polyamide, is a type of material made from Kevlar® (and other polyamides introduced in other varieties such as Twaron and Technora) – a material that has been called a superhero of fibres. It was invented in 1965 and has been used for such essential activities as winching cables on helicopters, mooring lines on ships and oil rigs, and body armour. Its super-strong and non-rust properties make it the perfect choice for a weapon like the Lasso of Truth.

Once we have this outer layer that is chemical resistant and occlusive, the next thing we need is an absorbent layer which will sit on top of this and let the truth serum drug permeate through the skin and to the bloodstream of the Lasso's captives. This will form a delivery vehicle for the truth serum with similar properties to a nicotine patch. This top layer will touch the captive's skin and so should be a selectively permeable membrane, allowing the truth serum to touch the skin (to ensure the effectiveness of the Lasso and that the safety of the captives is not compromised). This 'patch-Lasso' must also deliver the truth serum drug at a controlled rate, as an overdose of thiopental serum could result in sending the captives to sleep or even death. The top layer of the patch-Lasso must also be made from a plastic material that is flexible enough to be applied and removed from skin without breaking or tearing.

There must also be an adhesive material on the patch-Lasso which will attach it to the skin of its captives. This adhesive should be water-resistant in case nervous captives sweat, and non-allergenic so that it does not irritate the skin and distract captives from the fact-finding missions of the Lasso! The adhesive must also be pressure sensitive and of medical-grade standard. The best option for the Lasso's adhesive would therefore be acrylate polymers, which fulfil all the aforementioned criteria. The effect of the adhesive would also explain the possibility for being caught in the Lasso's power while only touching the rope. If we solve all of these design challenges, we may well have a fully functioning Lasso of Truth, but we may need to enhance ourselves morally to be effective users of it!

13.10   THE LASSO AND DEATH

While the Magic Lasso has been an extension of loving justice through the ages of the Wonder Woman franchise, there is also a darker side to this fascinating instrument. The power of the truth can be very dangerous and, in various episodes of Wonder Woman, the truth has led to deadly consequences. The most important example of this comes when Wonder Woman captures Maxwell Lord. Maxwell Lord is a supervillain who had been tormenting Superman's mind and forcing him into destructive acts, until he was captured by Wonder Woman, who made him confess how to stop his mind control of Superman. Maxwell Lord responded that she must kill him in order to stop the curse and, in her anger, Wonder Woman snapped his neck, thus breaking the Justice League code. Being a member of the Justice League (the TV series of which first aired in 2001), Wonder Woman is bound by the Justice League code, which stipulates that killing is prohibited for all members. Wonder Woman is, so far, the only member of the Justice League who has broken this code and it cost her dearly in reputation, status and existential angst after the murder took place. Therefore, we are shown that the Lasso of Truth is not an instrument to be used lightly as a tool for fun, but should instead be protected from unscrupulous use and treated with extreme caution.

REFERENCES

1. G. Johns(w), J. Lee(p) and G. Ha(a), *"The Villain's Journey"*, *Justice League #2*, DC Comics, February 2013.
2. W. M. Marston(w) and H. G. Peter(i), *The Wonder Woman Chronicles #2*, DC Comics, December 2011.
3. G. Simone(w) and B. Chang(p,i), *"Warkiller" Wonder Woman*, vol. 3, #37–41, DC Comics, December 2009.
4. G. Johns(w) and I. Reis(p), *"Blackest Night" Wonder Woman #1*, DC Comics, January 2010.
5. M. Jay, *The Virtues of Mendacity*, University of Virginia Press, Charlottesville, 1st edn, 2010.
6. D. Philips, *Wonder Woman #16 Review–IGN*, IGN, 2017 [cited 8 February 2017], available from: http://uk.ign.com/articles/2008/01/24/wonder-woman-16-review.
7. G. Rucka(w) and J. Lee(p), *"Sacrifice concludes" Wonder Woman*, #219, DC Comics, September 2005.
8. E. Hunt, *Wonder Woman Writer Confirms Superhero is Queer*, The Guardian, 2017 [cited 8 November 2016], available from: https://www.theguardian.com/books/2016/sep/30/wonder-woman-writer-confirms-superhero-is-queer.
9. *Stand Up for the Empowerment of Women and Girls Everywhere*, 2017 [cited 8 November 2016], available from: http://www.un.org/susatainabledevelopment/wonderwoman/.
10. J. Lepore, *The Last Amazon: Wonder Woman Returns*, The New Yorker, 2014.
11. S. M. Crothers, *"Meditations on Votes for Women"*, *The Atlantic, (July 1914)*, The Atlantic, 1914 [cited 8 November 2016], available from: https://www.theatlantic.com/magazine/archive/1914/07/meditations-on-votes-for-women/303358/.
12. M. Lamb, *Who Was Wonder Woman 1?* Bostonia, 2001 [cited 8 November 2016], available from: https://web.archive.org/web/20070104060542/http://www.bu.edu/alumni/bostonia/2001/fall/wonderwoman/.
13. B. Patton, Wonder Woman: bondage and feminism in the Marston/Peter comics, 1941–1948, by Noah Berlatsky, *J. Graphic Novels Comics*, 2015, 7(2), 223–224.
14. N. Berlatsky, *Wonder Woman's Violent, Man-Pandering Second Act*, The Atlantic, 2013 [cited 8 November 2016], available from: http://www.theatlantic.com/sexes/archive/2013/02/wonder-womans-violent-man-pandering-second-act/272871/.

15. J. Lepore, On Evidence: Proving Frye as a Matter of Law, Science, and History, *Yale Law J.*, 2015 [cited 8 November 2016], **124**(4), available from: http://www.yalelawjournal.org/article/on-evidence-proving-frye-as-a-matter-of-law-science-and-history.
16. M. Wagner(w), *"Trinity" #98 Part 1*, DC Comics, 1999.
17. M. Waid(w), B. Hitch(p) and P. Neary(i), *JLA*, vol. 1, #49, DC Comics, January 2001.
18. G. Scott, *A History of Torture*, 1st edn, Senate, 1995.
19. B. D. Bartholow, E. A. Henry, S. A. Lust, J. S. Saults and P. K. Wood, Alcohol effects on performance monitoring and adjustment: Affect modulation and impairment of evaluative cognitive control, *Journal of Abnormal Psychology*, 2012, **121**(1), 173–186.
20. CommunistCrimes.org - 1947–1965, Communistcrimes.org, 2016 [cited 8 November 2016], available from: http://www.communistcrimes.org/en/Database/Romania/Historical-Overview/1947-1965.
21. G. McEvoy, *American Hospital Formulary Service–Drug Information*, American Society of Health-System Pharmacists, Bethesda, 1st edn, 2003.
22. G. Edelman and V. Mountcastle, *The Mindful Brain*, MIT Press, Cambridge, 1st edn, 1978.
23. A. Karim, M. Schneider, M. Lotze, R. Veit, P. Sauseng and C. Braun, *et al.*, The truth about lying: inhibition of the anterior prefrontal cortex improves deceptive behavior, *Cereb. Cortex*, 2009, **20**(1), 205–213.

CHAPTER 14

Super Frequently Asked Questions

KARL BYRNE

The Royal Institution of Great Britain, 21 Albemarle Street, London, W1S 4BS UK
E-mail: karl.byrne@gmail.com

14.1 INTRODUCTION

I love comics. I know that might not come as much of a surprise, given this book's focus, but I thought I should make it clear at the start of the chapter. We've covered a lot of amazing superhero science in the book so far, but by no means everything! There are still lots of questions, queries and conundrums that didn't quite fit into the previous chapters. So, this chapter will attempt to provide scientific explanations for, well, everything else ... However, I must accept that in some cases the answer will have to be 'magic' or 'that's how the comic book universe works', and sometimes that's OK. In some cases, I try to explain something as if it's happening in the real world, while in others I've attempted to explain them in the comic universe, or "in-universe", in nerd speak. Hopefully I've made it clear when I'm talking about the real world and when I'm talking within the comic universe. Without further ado, on to the questions.

The Secret Science of Superheroes
Edited by Mark Lorch and Andy Miah
© The Royal Society of Chemistry, 2017
Published by the Royal Society of Chemistry, www.rsc.org

14.2   IS IT POSSIBLE TO CLIMB UP A WALL LIKE SPIDER-MAN?

When teenager Peter Parker gets bitten by an irradiated spider, he 'acquires the agility and proportionate strength of an arachnid'.[1] This makes him incredibly strong, gives him amazing reflexes, a 'spider sense' and, of course, the ability to climb up buildings. But could we actually do this in real life? Could we devise some kind of body suit or extra sensory capacity to mimic our favourite wall-crawler's ability, in the real world? Zoologists at Cambridge don't think so. In December 2015 they concluded that the ability to climb up walls would forever remain outside of our grip[2] and to understand why, we have to examine what's actually going on with a spider when it crawls up walls and along ceilings.

The ability to climb walls has evolved independently a number of times and can be observed in various species besides spiders, such as geckos. Both spiders and geckos use the same mechanism to defy gravity and, in fact, geckos are much better at doing it than spiders. The mechanism was discovered in the 2003 by biologist Kellar Autumn, who was inspired when a spider nearly dropped on his head, having been chased by a gecko.[3] After getting over the fright of their spider close encounter, the arachnophobic Autumn spent the next 6 years looking for the answer.

The secret to the sticky pads on both gecko and spider feet is due to their specialized hairs, called setae. These hairs branch smaller and smaller bristles until they are between 10 and 30 nm thick, or roughly 1000 times thinner than a human hair. At this size (1/100 000 000 of a metre) these tiny bristles, called spatulae, can utilize something called van der Waals forces.[4] These forces come into effect when the electrons in the molecules of the spatulae interact with the electrons in molecules on the wall or ceiling, thus creating electromagnetic attraction. Although each spatulae creates only a tiny attractive force, their vast number means that geckos can hold many times their own weight. Furthermore, as long as the gecko keeps its feet clean (and they spend a lot of time licking their feet to do just this), the stickiness will never wear out, allowing them to stick to practically any surface, apart from Teflon®. Spiders use this same phenomenon to accomplish their wall crawling, but spiders weigh a lot less and have 8 legs, which, in my opinion, is cheating. This means that gecko girl would be a much better climber than Spider-Man. It also means that that Spider-Man has very hairy palms.

So what's the problem? If spiders and geckos can do this, why is it so difficult for a human to accomplish the same effect? The Cambridge paper hinges on an old problem: Surface area does not increase at the same rate as volume. In other words, as you get bigger, your weight increases and you would need much more surface area for more setae, to continue to have the ability to sustain your weight. For instance, a house spider uses up to 0.9% of its surface area for setae, while a gecko uses about 4–4.5% of its total surface area. If we scale up this ratio to human size, then we would need to have about 40% of our body covered in setae – and they would all have to be on the same side of your body. Either that, or you would need to have very large hands and feet.

But wait! There's another part to the story … Although the Cambridge team published a paper in 2015 saying that it was impossible, 2 years earlier Elliot Hawkes, an engineer from Stanford University, showed the exact opposite, by making his very own gecko gloves and scaling a 3 metre high glass wall.[5] Each glove had 24 postage stamp-sized tiles of gecko-inspired sticky material, allowing the force to be evenly distributed across the pads, which allowed him to climb, albeit in a slightly awkward looking way. However, he still didn't have Spider-Man's strength and agility, meaning web-crawling remains awkward-looking and very tiring! Although we may not see scores of spider-men climbing buildings in the near future, the materials have a lot of other uses, from stitch free wound repair to helping robotic arms in space catch satellites and space debris.

14.3 IF YOU CUT WOLVERINE IN HALF, WOULD HE GROW BACK AS TWO WOLVERINES?

James Howlett, the short, hairy Canadian mutant more commonly known as Logan, or by his codename Wolverine, is well over 100 years old, an age he enjoys due to his mutant healing ability. This regeneration superpower means that he ages incredibly slowly and can recover from pretty much any injury. As well as his ability to heal, he also has heightened senses, retractable claws and some major anger management issues. One of the most skilled fighters in the Marvel universe, he is a force to be reckoned with and that's before you consider that his entire skeleton has become intertwined with adamantium; a near indestructible metal alloy. In his very first appearance he battled the Hulk[6] and, although he didn't beat him, they were very well matched.

Knowing all this means that your chances of chopping Wolverine in half are not going to be good, which may make our question more of a philosophical nature, rather than a scientific one. However, there are a few weapons in existence that could cut through Wolverine's skeleton, most notably the Muramasa blade, designed specifically to be used against him.[7] So, assuming you have the skills and tools to cut Wolverine in half, what would happen?

Let's have a look at the animal kingdom and see if we can find some answers. There are quite a few animals that have the ability to regenerate limbs.[8] Certain species of lizards (like Anolis) can lose their tails as a defence mechanism, so that hungry predators get a tasty mouthful, but not the whole animal. It takes a while for the tail to grow back, and it's usually smaller than the original one, with cartilage instead of bone, but it's better to lose a tail than to get eaten.

Salamanders and newts go even further. They have the ability to regenerate entire limbs over and over again and they can also replace vital organs, even their brains! Starfish and other echinoderms can regenerate limbs and can use this as a more extreme form of defence, where they can detach a limb, which will crawl off on its own to distract a predator. The starfish can then regrow the missing limb. A few starfish are even better at the regeneration game – as well as the original starfish re-growing the detached limb, the detached limb can go on to grow into a complete starfish.

So, is Wolverine more like a lizard, a newt or a starfish? Looking through the comics, he has been reduced to a skeleton and still managed to recover, which shows he has an incredible ability to heal. However, at one point he got ripped in half by the Hulk and, on this occasion, both halves didn't regrow into new Wolverines. In fact, he had to crawl up Mount Everest to find his legs and put himself back together[9] (the pedant in me should note that this was in the Ultimate Universe, where his powers differ slightly from the Wolverine we usually see in the comics and movies).

Wolverine has been in a fair few fights and has had chunks lopped off, which (as far as we know) haven't gone on to grow into new Wolverines. So, if you did cut him exactly in half there's a chance you could end up with two Wolverines, each with half a skeleton coated in adamantium and with different memories and personalities due to the way the brain is divided up. Alternatively, it might be that the legs just wither and the part of the body which has the vital organs would then grow new legs. Of course, our answer might also vary depending on which way he is cut – top to bottom, or left to right! However, what seems more likely to happen is that the two halves will get back together. The result is that you'll end up facing a very, very angry Wolverine who will want to have a rather intense conversation with you, involving claws.

14.4   IS THERE A LINK BETWEEN SUPERPOWERS AND HAVING AN ALLITERATIVE NAME?

Peter Parker, Clark Kent, Bruce Banner, Wally West, Reed Richards ... The list goes on and at first glance one might think that having an alliterative name makes it more likely to develop superpowers. Or is it just because the creators of comic book universes have a more poetic nature? I'm going to argue that this is the case. Let me explain.

If we divide the superheroes and villains into the two biggest comic book universes of DC and Marvel, it becomes clear very quickly that the Marvel Universe is populated with a lot more alliterative names than the DC universe, with the notable exception that nearly every person that Superman knows has the initials L L (Lois Lane, Lana Lang, Lex Luthor, Lionel Luthor, Lyla Lerrol and Liri Lee, to name but a few – this pattern goes right back to the earliest days of Superman in Action Comics, and might have started as a reference to the initials of Superman co-creator Joe Schuster's girlfriend whose initials, L L, have been used as a running joke by writers ever since). However, there is a further reason why Marvel is so alliterative crazy and that reason is the Man, the Myth, the Legend that is Stan Lee.

Lee has been quoted as saying that he used so many alliterative names because they 'sounded more fun'[10] and what he is alluding to may have a scientific explanation. This is because our brains love patterns, and alliteration and rhymes let our brains spot simple patterns. The joy of making these rapid observations of a pattern gives us a reward and makes us enjoy them better, making them sound more fun. This same mechanism is also linked to why we enjoy rhythmic music and may also help explain why we want to dance when we listen to music, and why dancing is so enjoyable.[11] By associating character names with pleasure, this means that we enjoy reading about them and become more invested in what happens to the characters, which may also explain the popularity of Spider-Man, Superman and the Hulk.

There is also a link between memory and alliteration, where things that rhyme are easier to remember.[12] This may also be a reason why Lee used alliteration so frequently. In the 1960s Stan Lee wrote nearly every Marvel comic on his own and needed to remember the plethora of different characters. By giving them first names and surnames with the same letter, this made it easier to remember all of the new characters he had created. However, this method wasn't fool proof. In Spider-Man's second ever appearance, he mistakenly calls his alter ego 'Peter Palmer'[13] and he's also called the Hulk's alter ego Bruce Banner, 'Bob' Banner!

14.5   IS IT POSSIBLE FOR A NORMAL PERSON TO BECOME BATGIRL?

The simple answer to this is 'yes'.

Over the years, there have been various versions of Batgirl, but the most famous is Barbara Gordon, daughter of Gotham's home-bred Commissioner Jim Gordon. Like Batman, she has no superpowers, instead using her skills, intelligence and gadgets to fight crime.

Batgirl is a skilled martial artist, having black belts in judo and karate, and is described as a star athlete. She has been training in self-defence since she was a small child and is an expert in multiple small arms, including Batman's Batarangs.[14] Batgirl is also an expert in computers and electronics, using her computer-hacking skills in her crime-fighting adventures. All of this is well within the realm of feasibility, if you're willing to put the work in and if you have the talent.

A slightly harder problem is that Barbara Gordon has a genius-level intellect and a photographic memory. However, even if you do not feel like you have these gifts, there are ways to improve your memory, including mnemonics, and the Method of Loci – a way of using locations to remember important facts, that dates back to at least 100 BCE.[15]

The final part of Batgirl's crime-fighting arsenal are the gadgets, some which she gets from Batman, others which she builds herself. So, all you need to become your very own Bat-vigilante is a lot of hard work, training and a bit of ingenuity in gadget creation.

14.6   WHY ARE SO MANY SUPERHEROES AND VILLAINS SCIENTISTS?

Not everyone with superpowers is a scientist, but a large proportion of them have been involved in science, technology, engineering or medicine: Tony Stark, Hank Pym, Reed Richards, Bruce Banner, Victor Von Doom, Norman Osborne and Hank McCoy, to name but a few. Two things that may explain this are their desire to investigate 'forces they don't understand' (which could describe the origins of the Fantastic Four, as they investigated cosmic rays that imbued them with their powers) and the desire to test their experiments on themselves. In the real world 'investigating forces they don't understand' could (very loosely ...) be said to explain why we have such amazing projects like the Large Hadron Collider and the search for dark matter (rather than creating superpowers!) and such investigations are an important part of the scientific process. The primary purpose of science is to reveal our underlying reality and help us understand the forces of nature.

The other common reason that explains why comic-book scientists tend to become super-powered individuals – that they test experiments on themselves – usually does not work out very well in the real world. More often than not, self-experimentation leads to a trip to hospital (or worse, the morgue) and there have been historical cases of scientists testing their experiments on themselves, with varying degrees of success (and/or fatality). Note that this practice is often considered to be deeply unethical and unscientific. Having said all of that, there have been instances where self-experimentation has led to remarkable discoveries.

One of the most famous cases of self-experimentation happened in 1984 when Australian scientist Barry Marshall purposely infected himself with the *Helicobacter pylori* bacterium. Marshall hypothesized that the cause of stomach ulcers was actually this bacteria; this view went against the conventional medical wisdom that stomach ulcers were caused by stress. Having failed to infect piglets with the bacteria, Marshall put a camera down his throat and took a biopsy, which showed that he had a healthy stomach. He then drank a dish of *H. pylori*, expecting to develop an ulcer within a year. Within a few days Marshall started feeling vague nausea and developed bad breath and a few days later started vomiting. Two weeks after drinking his bug-filled drink, he had a second endoscopic investigation and biopsy which showed massive inflammation. On taking antibiotics the symptoms quickly cleared up. His work was integral to proving that the ulcers had a bacterial cause and led to new treatments being developed to treat gastric ulcers. Although Barry Marshall didn't develop any superpowers from his culture cocktail, he was awarded the Nobel Prize in Physiology or Medicine in 2005, so it wasn't all bad! So, it's possible that being a superhero inherently involves having some of the qualities exhibited by scientists, not least because both have the capacity to take existing materials and knowledge, and create something completely new.

14.7   HOW DOES SUPERMAN HAVE HIS HAIRCUT?

Kal-EL was sent to earth by his parents, escaping his doomed home planet of Krypton. Landing in Kansas and being adopted by Jonathan and Martha Kent, he grew up to become mild-mannered reporter Clark Kent. However, his alien physiology was energized by Earth's yellow sun and he developed incredible superpowers. Deciding to use them for the benefit of humanity, he became Superman.

Superman is known for his vast array of powers, which include being 'faster than a speeding bullet, more powerful than a locomotive.' Moreover, his powers have developed over time, so that he has X-ray vision, invulnerability, super strength, super speed, super hearing and, at one point, even super hypnotism. While it may seem as an incredible advantage, invulnerability has the downside that personal grooming becomes pretty difficult. How do you cut invulnerable hair?

There is a great 1946 comic cover featuring a poor barber destroying garden shears trying to cut Superman's hair[16] and another is found in 1968, depicting Jimmy Olsen trying to do the same.[17] However, the answer as to how it is possible is found in another comic from 1986, in which writer John Byrne re-tells Superman's origin story.[18] After Lois Lane tells Clark to go and shave his face, Clark goes into the bathroom and uses his heat vision, bouncing it off a fragment from the rocket in which he crash landed to earth as a baby. As the rocket is one of the only materials strong enough to survive a blast from his heat vision, the rebound onto his face allows him to use the heat to shave his face. Simple! His cousin Supergirl also has the same trouble and uses a similar solution – shaving her legs with her heat vision.

Other near indestructible characters have a harder time, especially if they don't have access to Kryptonian X-ray or heat vision. For instance, Marvel character She-Hulk has asked for an adamantium razor to be made for her so she can shave her legs![19] Alternatively, Wolverine's problem isn't that he can't shave, it's just that his mutant healing factor makes the hair grow back within a couple of hours.

If super vision wasn't an option, perhaps there is still a scientific solution. With recently developed super-hard materials, Superman might find a simpler solution that involved grinding away his face fuzz. Hyperdiamonds may provide an answer. Hyperdiamonds are similar to regular diamonds, but are about twice as strong and have 'superior wear resistance'.[20] This could allow Superman to grind off his hair before he completely destroys the super-sandpaper.

14.8   WHY DOESN'T MAGNETO EVER NEED A TOWEL?

Magneto is awesome! His ability to control magnetic fields means that he can control much more of his environment than you might first imagine. When we think of magnetic things, metallic objects containing iron (and to lesser extents cobalt and nickel) come to mind. After all, we're pretty familiar with magnets sticking together in the real world. Our school physics lessons may not have covered much more than this form of 'ferromagnetism', but there are others, the most notable of which are paramagnetism and diamagnetism and these other forms explain just how powerful Magneto really is.

Paramagnets are attracted to magnetic fields, whilst diamagnetic are repelled by them, but, unlike ferromagnets, neither generate their own external magnetic fields. Within the paramagnetic camp there's tungsten, caesium and even oxygen. Meanwhile, hydrogen is the most potent diamagnet, even though the force of the diamagnetic interaction between hydrogen (and common molecules that contain it, I'm thinking water here) is very weak. However, with the right equipment you can see diamagnetism taking place in real life. Professor Mark Lorch provides a nice demonstration of this using a couple of cherry tomatoes stuck on a wooden kebab stick, which are then balanced the point of a pin.[21] When a strong magnet is placed near the tomatoes, they are repelled by the force and the contraption starts to spin, demonstrating the principle of diamagnetism.

Since Magneto has the capacity to generate vast magnetic fields he'll easily be able to move water about. So, once he steps out of the shower, all he needs to is invoke his marvellous powers to repel the water and hey presto, a bone dry supervillain.

14.9 WHY DOESN'T FROZONE'S HAND FREEZE?

You will recall that Frozone is one of the characters from the 2004 film *The Incredibles* and he has the capacity to create ice, but why doesn't his own hand freeze in the process of generating such frozen goods? Somehow, he seems able to drop the temperature of his hands, causing water vapour to condense out of the air and then form it into all sorts of icy structures. But when he performs this feat why doesn't his hand freeze along with the condensed water? It could be that he is using a trick that has a biological precedent in fish, plants and insects that live in sub-zero conditions.

We all know that water starts to freeze at below 0 °C, but it is actually quite easy to chill really clean water (a bottle of mineral water will do) well below that point without it solidifying. This super-cooled water stays liquid because ice crystals need somewhere to grow from; usually bits of dust or debris work nicely as nucleation sites, but other ice crystals work even better.

Fish that live in polar regions have a way of disrupting ice crystal growth with anti-freeze proteins. These proteins bind to the corners of ice crystal from which they grow and stop other water molecules from attaching themselves. This stops the expansion of the crystal in its tracks and keeps the super-cooled water fluid. So, perhaps this is what's happening with Frozone.

14.10 CONCLUSIONS

This concludes our rapid fire round-up of all the remaining questions. While there are hundred of other characters who we have yet to consider and an even larger number of superheroes that we have yet to imagine, one thing is clear. The science behind superheroes is not entirely outside of the realms of the physical possibilities that operate within our world. In fact, some of the precedents for discovering superhuman powers involve simply looking to other sources of nature, where such capacities exist. In this respect, the most amazing thing about a superhero is that their powers are much closer to our world than we might have thought initially and this is a very humbling realization.

REFERENCES

1. S. Lee and S. Ditko, *Amazing Fantasy #15*, Marvel Comics, New York, N.Y., 1962.
2. D. Labonte, C. Clemente, A. Dittrich, C. Kuo, A. Crosby and D. Irschick, *et al.*, Extreme positive allometry of animal adhesive pads and the size limits of adhesion-based climbing, *Proc. Natl. Acad. Sci.*, 2016, **113**(5), 1297–1302.
3. E. Sohn, *How a Gecko Defies Gravity*, Science News for Students, 2017 [cited 8 February 2017], available from: https://www.sciencenewsforstudents.org/article/how-gecko-defies-gravity.
4. K. Autumn, M. Sitti, Y. Liang, A. Peattie, W. Hansen and S. Sponberg, *et al.*, Evidence for van der Waals adhesion in gecko setae, *Proc. Natl. Acad. Sci.*, 2002, **99**(19), 12252–12256.
5. R. Lawler, *Stanford's 'Gecko Glove' Makes Spider-Man Climbing Possible*, Engadget, 2017 [cited 8 November 2016], available from: https://www.engadget.com/2016/01/28/stanfords-gecko-glove-makes-spider-man-climbing-possible.
6. E. E. Ruppert, R. S. Fox and R. D. Barnes, *Invertebrate Zoology*, Thomson Books, Canada, 1st edn, 2006.
7. L. Wein and J. Romita, *Incredible Hulk Vol. 1 #181*, Marvel Comics, New York, N.Y., 1974.

8. C. Claremont and J. Buschema, *Wolverine Vol. 2 #2*, Marvel Comics, New York, N.Y., 1988.

9. D. Lindelof, *Ultimate Comics Wolverine vs Hulk*, Marvel Enterprises, 1st edn, 2010.

10. S. Lee and J. McLaughlin, *Stan Lee*, University Press of Mississippi, Jackson, 1st edn, 2007.

11. S. Brown and L. Parsons, The neuroscience of dance, *Sci. Am.*, 2008 [cited 8 November 2016], **299**, 78–83, available from: http://www.neuroarts.org/pdf/SciAm_Dance.pdf.

12. R. Lea, D. Rapp, A. Elfenbein, A. Mitchel and R. Romine, Sweet silent thought: alliteration and resonance in poetry comprehension, *Psychol. Sci.*, 2008, **19**(7), 709–716.

13. S. Lee and S. Ditko, *The Amazing Spider-Man #1*, Marvel Comics, New York, 1963.

14. B. Randal, *Secret Origins # 20 Batgirl and the Golden Age Dr Mid-Nite*, DC Comics, 1987.

15. J. O'Keefe and L. Nadel, *The Hippocampus as a Cognitive Map*, Clarendon Press, Oxford, 1978.

16. D. Cameron and S. Citron, *Superman #38*, DC Comics, 1946.

17. J. Shooter and C. Swan, *Superman's Pal, Jimmy Olsen #110*, DC Comics, 1968.

18. J. Byrne, *Superman: The Man of Steel*, DC Comics, 1986.

19. C. Soule and J. Pulido, *She-Hulk #2*, Marvel Comics, 2014.

20. N. Dubrovinskaia, S. Dub and L. Dubrovinsky, Superior wear resistance of aggregated diamond nanorods, *Nano Lett.*, 2006, **6**(4), 824–826.

21. M. Lorch, *Magnetic Tomatoes*, 2014 [cited 8 November 2016], available from: https://youtu.be/yMbENMTrIGY.

Subject Index

References to tables and charts are in **bold** type

oat cakes, 1
octopus, 151
  mimic octopus
  *(Thaumoctopus
  mimicus)*, 28
oil rigs, 193
Olsen, Jimmy, 204
Olympic athletes, 1–3, 7,
  72–3
Olympus, 184
oncogenes, 40
oobleck recipe, 114
Operation Kronos, 152
Osborne, Norman, 203
oscillating electric field, 135
osteocalcium phosphate,
  152
Osterman, Jon (Dr. Manhattan),
  24, 83, 85–6, 91
outbreaks, 51–4, 57, 59, 62
ovalbumin, 9, **10**
owls, 23–4
oxidative phosphorylation
  (OXPHOS), 43
oxygen, 5–6, 23, 37, 41–6, 76,
  100, 133, 206
oxygenase-1, 45
oxytocin, 44, 165

palm repulsors, 124–5
Pankhurst, Emmeline, 185
panthers *(Puma concolor
  coryi)*, 21
Paradise Island, 184–5
paraffin, 121
paramagnetism, 205
paranoid schizophrenic, 174
Parker, Peter (Spider-Man),
  1, 7–9, 12–3, 15, 19, 26,
  149–50, 198–9, 201–2

Parr, Helen neè Truax
  (Elastigirl), 98, 151–3,
  156–7, **158**, 160–5
particle colliders, 83, 86
passive oxide, 100
patch mechanism, 192–4
patient zero, 62
Patton, George, 69
PayPal®, 179
peanut butter sandwiches, 5, 7
Peepers, 24
pelvis, 152, 164
perfluorocarbons (PFC), 43
periodic table endangered
  elements, **84**
petalite, 87–8
Pheidole morrisi
  (soldier ants), 73
phenotype, 20–2
phenylalanine, **10**
phosphorescent screen, 72
phosphorus, 89
photographic memory, 202
photoreceptors, 23–4, 132
photosynthesis, 147
pilots, 78–9, 142, 151, 173
pit organs, 25
pitohui birds *(Pitohui
  dichrous)*, 29
Pitohui dichrous (pitohui
  birds), 29
Pixar, 152
placebo, 64
plague, 52
plankton, 4
plasma, 6
plastic deformation, 98–9,
  108, 164
plastics, 98–9, 108, 124, 155,
  163, 164, 193

superhydrophilic polymers, 164

Superman (Kent, Clark), 24–5, 112, 121, 142, 177, 184–6, 194, 201–5

surgery, 56, 71, 79, 123

Sweden, 87–8

swiflets, 25

Switzerland, 72

symphysis pubis dysfunction (SPD), 165

Syndrome, 152, 163

synthetic blood, 43

Système international d'unités (SI), 147–8

systolic blood pressure test, 187–8, 190

Szostak, Jack, 40

T'Challa, 95

tablet, 86, 89

Tactical Assault Light Operator Suit (TALOS), 127

tanks, 105, 172

tapetum, 134

Technora, 193

Teenage Mutant Ninja Turtles, 19

Teflon®, 198

telomerase, 40

telomeres, 40

temperature-dependent sex determination, 22–3

tennis, **3**

tensile strength, 8, 12, 96, 98–101, 113–5

Terminator's Skynet, 168

Tesla coils, 117–8

Tesla, Nikola, 118

Thaumoctopus mimicus (mimic octopus), 28

The American Scholar, 187

The Dead, 63

The Enemy, 63

The Fantastic Voyage, 167

The Fear, 63

The Flash (Allen, Barry), 141–8

The Girl with all the Gifts, 58, 63

The Green Lantern, 184

The Hitch Hiker's Guide to the Galaxy, 175

The Hot Zone, 58

The Hulk, 15, 19, 33–47, 80, 126, 200, 202, 205
   see also Banner, Bruce

The Human Torch (Storm, Johnny), 6–7

The Incredibles, 19, 163, 206

The Invisible Man (Griffin), 132–5

The Invisible Woman (Storm, Susan), 130–1, 134–5, 138

The Myth of Superman, 186

The Thing
   see Grimm, Ben

Themyscira, 185

theory of relativity, 135

thermal conductivity, 7

thermal imaging, 72

thiopental serum, 192–3

Third Reich, 70

Thor, 80, 96, 118

Three Laws of Robotics, 176

three-point-bend test, 113

threonine, 9, **10, 12**

Through the Looking-Glass, 21

time travel, 146